PROTEIN BIOCHEMISTRY, SYNTHESIS, STRUCTURE AND CELLULAR FUNCTIONS

HUMAN SERUM ALBUMIN (HSA)

FUNCTIONAL STRUCTURE, SYNTHESIS AND THERAPEUTIC USES

PROTEIN BIOCHEMISTRY, SYNTHESIS, STRUCTURE AND CELLULAR FUNCTIONS

Additional books in this series can be found on Nova's website
under the Series tab.

Additional e-books in this series can be found on Nova's website
under the e-book tab.

PROTEIN BIOCHEMISTRY, SYNTHESIS, STRUCTURE AND CELLULAR FUNCTIONS

HUMAN SERUM ALBUMIN (HSA)

FUNCTIONAL STRUCTURE, SYNTHESIS AND THERAPEUTIC USES

TRAVIS STOKES
EDITOR

New York

Library of Congress Cataloging-in-Publication Data

ISBN: 978-1-63482-963-2
Library of Congress Control Number: 2015940245

Published by Nova Science Publishers, Inc. † New York

CONTENTS

PREFACE

Human Serum Albumin (HSA) is the most abundant plasma protein. It has been widely used for drug delivery systems and has recently emerged as a versatile carrier for therapeutic agents against diabetes, cancer and infectious diseases. This book provides an overview of the expanding field of preclinical and clinical applications and developments that use albumin as a carrier of drug delivery systems. The authors' discuss the properties of drug binding sites within the structure of HSA, discuss new possibilities for the therapeutic potential of HSA and analyze recently reported HSA-drug complexes including HSA-antibody conjugates. Novel investigations on the applications of albumin fusion proteins are discussed as well, with a focus on tumor targeting and intracellular delivery. Other chapters examine the different aspects of albumin glycation and oxidation, the changes in the structure of human serum albumin determined from infrared spectroscopy and a review of CAPIDAN, a special fluorescent dye, which attaches to drug binding sites of human serum albumin.

Chapter 1 – Human serum albumin has recently emerged as a versatile carrier for therapeutic agents against diabetes, cancer and infectious diseases. Market approved products include fatty acid derivatives of human insulin or the glucagon-like-1 peptide (Levemir®) for the treatment of diabetes, a long-acting injectable human interferon alpha2b fusion protein (Albuferon®) for the potential treatment of hepatitis C virus infections, and the taxol albumin nanoparticle Abraxane® for the treatment of metastatic breast cancer, advanced non-small cell lung cancer and advanced pancreatic cancer. In addition, an increasing number of albumin-based drugs are currently in clinical trials. These include an antibody fusion protein for treating HER2/neu positive breast cancer, and a recombinant fusion protein linking factor IX with albumin for the Prophylaxis of hemophilia B. In the preclinical setting, next generation approaches include albumin binding bioactive gases (bio-Gas) such as nitric oxide or hydrogen sulfide (*S*-Nitrosated Albumin analogs and Albumin Persulfide) for treating ischemic/reperfusion injury, cancer and bacterial infections. This review provides an overview of the expanding field of preclinical and clinical applications and developments that use albumin as a carrier of drug delivery systems.

Chapter 2 – Human Serum Albumin (HSA) has been widely used for drug delivery systems. HSA binds to variety of drugs resulting in HSA-drug complexes. HSA influences the pharmacokinetic and pharmacodynamic properties of drugs and the complexes have demonstrated potential for improving drug tracking and efficiency. This review discusses the properties of drug binding sites within the structure of HSA. The authors analyzed recently reported HSA-drug complexes including novel HSA-antibody conjugates. According to the literature reviewed, they present an overview of the HSA-drug complexes that may potentially lead to significant clinical applications by improving drug efficiency and control of distribution and release rate of drug. Moreover, the authors discuss the fabrication of HSA drug delivery systems and the factors that optimize particle sizes and distribution and the effect of pH on the HSA-drug complexes. Also, the authors discuss delivery of HSA-anticancer drug complexes to receptors expressed on the cancer cells surfaces. Research concerning *in vitro* and *in vivo* techniques used for the HSA-drug complex monitoring is also presented in this chapter.

Chapter 3 – Human serum albumin (HSA) is the most abundant plasma protein. It has been extensively studied and applied in medicine and pharmaceutics. Taking advantage of the recombinant DNA technology, fusion proteins with novel functions have been generated. The recent approval of the albumin-GLP-1 fusion protein drug, Tanzeum, will encourage researchers to explore many more advantages of HSA for therapeutic applications. Generally, these albumin fusion proteins possess long serum half-life, good solubility, and versatile ligand binding ability. In addition to being used as drugs, albumin fusion proteins can serve as carriers for drug delivery to improve drug efficacy and safety. Herein, novel investigations on the applications of albumin fusion proteins will be discussed, with focus on tumor targeting and intracellular delivery. These approaches carry the potential to co-deliver multiple drugs to the same cellular target for synergistic therapeutic effects.

Chapter 4 – Human serum albumin (HSA) is synthesized predominantly in the liver, and has an extraordinarily long circulatory half-life (~19 days). Since HSA is responsible for 80% of the colloidal osmotic pressure of plasma, the patients with hypoalbuminemia are typically given a plasma expander to increase the plasma volume. Clinically, using HSA as an extender is preferable to other options, such as dextran, hydroxylethyl starch, due to its long retention in plasma. However, the circulatory half-life of HSA is significantly-impacted by structural changes of HSA and pathological conditions. Recent scientific advances have revealed new information regarding the factors that influence the pharmacokinetic properties of HSA. In this chapter, the authors provide an overview of the impact of HSA structure and biotransformation upon the pharmacokinetic properties of HSA, and discuss new possibilities for the therapeutic potential of HSA based on its pharmacokinetic properties.

Chapter 5 – Glycative and oxidative stress are frequently observed in numerous disease states. From recent studies, it has become evident that protein glycation and oxidation play major roles in the activity, unfolding, and degradation of proteins, as well as for cell function. Human serum albumin (HSA), the major protein in the blood circulation, frequently undergoes increased glycation and oxidation in major diseases such as diabetes and renal failure. This chapter examines the different aspects of albumin glycation and oxidation by focusing on the impact of these processes on the structural, biological and physiological modification of the protein. Further, the importance of glycated and oxidized albumin as a biological marker for many diseases such as diabetes and renal failure is discussed.

Chapter 6 – The effect of organic solvent on the hydration and structure of human serum albumin was characterized using infrared spectroscopy in the thermodynamic water activity range from 0 to 0.98 at 25°C. Dioxane was used as a model organic solvent. This organic solvent may be considered as an informative molecular probe for analyzing the effect of hydrogen bonding and hydrophobic interactions on the hydration and structure of proteins.

The obtained results show that the hydration and structure of human serum albumin depend markedly on how the protein has been hydrated – whether in the presence or in the absence of organic solvent. Two organic solvent effects on water sorption by human serum albumin were observed: At low water activity (a_w<0.5), the water adsorption branch obtained in the presence of organic solvent is similar to that measured in the absence of dioxane. At high water activity (a_w>0.5), the water sorption isotherm in the presence of dioxane lies above the corresponding isotherm for pure water. This result corresponds to the organic solvent – assisted effect on water binding by human serum albumin. Changes in the structure of human serum albumin were determined from infrared spectra by analyzing the structure of amide I band. It was found that, at low water activity, the protein-protein contacts in the dried protein largely govern its thermodynamic and structural properties. At high water activity, the protein state is determined by the protein-organic solvent and protein-water interactions. The results from the thermodynamic and structural measurements were analysed to give a unified picture of the hydration process in the absence and presence of organic solvent.

Chapter 7 –CAPIDAN, a special fluorescent dye, binds to drug binding sites of human serum albumin. About 90-95 per cent of its fluorescence in serum or plasma originates from albumin-bound dye. This specificity opens a new way to study albumin in unfractionated serum saving intact albumin structure. Due to these properties, CAPIDAN has shown that a variety of diseases is accompanied by some changes of the drug-binding sites. This is a new source of information useful for clinics. That sensitivity to site abnormalities is raised by using time-resolved measurements of fluorescence decay and three new methods of the data analysis: (1) estimation of absolute values of concentration of fluorescent species in heterogeneous biological objects; (2) measurement of dye rotational mobility in heterogeneous albumin binding sites; and (3) method of chloride/nitrate substitution to detect interaction of a drug negative charge with arginine positive charges in the albumin binding sites.

In: Human Serum Albumin (HSA)
Editor: Travis Stokes

ISBN: 978-1-63482-963-2

Chapter 1

POTENTIAL USE OF HUMAN SERUM ALBUMIN AS A DRUG DELIVERY SYSTEMS

Yu Ishima[1], Toru Maruyama[1] and Masaki Otagiri[2,3,*]
[1]Department of Biopharmaceutics, Graduate School of Pharmaceutical Sciences, Kumamoto University, Kumamoto, Japan
[2]School of Pharmacy, Faculty of Health Sciences, Curtin Health Innovation Research Institute, Curtin University, Perth, Western Australia, Australia
[3]DDS Research Institute, Sojo University, Kumamoto, Japan

ABSTRACT

Human serum albumin has recently emerged as a versatile carrier for therapeutic agents against diabetes, cancer and infectious diseases. Market approved products include fatty acid derivatives of human insulin or the glucagon-like-1 peptide (Levemir®) for the treatment of diabetes, a long-acting injectable human interferon alpha2b fusion protein (Albuferon®) for the potential treatment of hepatitis C virus infections, and the taxol albumin nanoparticle Abraxane® for the treatment of metastatic breast cancer, advanced non-small cell lung cancer and advanced pancreatic cancer. In addition, an increasing number of albumin-based drugs are currently in clinical trials. These include an antibody fusion protein for treating HER2/neu positive breast cancer, and a recombinant fusion protein linking factor IX with albumin for the Prophylaxis of hemophilia B. In the preclinical setting, next generation approaches include albumin binding bioactive gases (bio-Gas) such as nitric oxide or hydrogen sulfide (*S*-Nitrosated Albumin analogs and Albumin Persulfide) for treating ischemic/reperfusion injury, cancer and bacterial infections. This review provides an overview of the expanding field of preclinical and clinical applications and developments that use albumin as a carrier of drug delivery systems.

[*] Corresponding author: Masaki Otagiri, PhD, Faculty of Pharmaceutical Sciences, Sojo University, 4-22-1 Ikeda, Nishi-ku, Kumamoto 862-0082, Japan. Phone.: +81-96-326-3887. Fax: +81-96-326-3887, E-mail: otagirim@ph.sojo-u.ac.jp.

1. INTRODUCTION

Among the available potential colloidal drug carrier systems, protein-based nanoparticles are particularly interesting because of their clear advantages that include greater stability during storage and *in vivo*, the fact that they are non-toxic and non-antigenic and the ease of scale up during their manufacture over other drug delivery systems [1-4]. Albumin is an attractive macromolecular carrier that has been shown to be biodegradable, nontoxic, metabolized *in vivo* to produce innocuous degradation products, non-immunogenic, easily tc purified and soluble in water allowing ease of delivery by injection, all of which make it an ideal candidate for nanoparticle preparation [5].

Albumin-based nanoparticle carrier systems represent an attractive strategy, since a significant amount of a drug can be incorporated into the particle matrix because of the different drug binding sites present on the albumin molecule [6]. The four main technologies that use albumin as a drug carrier are summarized in Figure 1. Drugs, prodrugs or polypeptides can either be bound physically or covalently through a ligand or a protein-binding group to Human serum albumin (HSA) (top left and right). More complex systems are characterized by the attachment of numerous targeting ligands and prodrugs that bind to the protein surface or nanobodies or bispecific antibodies that bind physically or are fused with albumin replacing the Fc fragment of immunoglobulin G (bottom right). Finally, nanoparticles, micellar structures or microbubbles that encapsulate lipophilic drugs or diagnostic agents can be prepared as water-soluble galenic formulations for suitable for intravenous injection (bottom left).

In the following chapters, we describe the recent developments in these different areas of applications that have progressed commercially, clinically as well as in preclinical settings and are continually expanding the field of albumin-based drug delivery concepts (Table 1, 2).

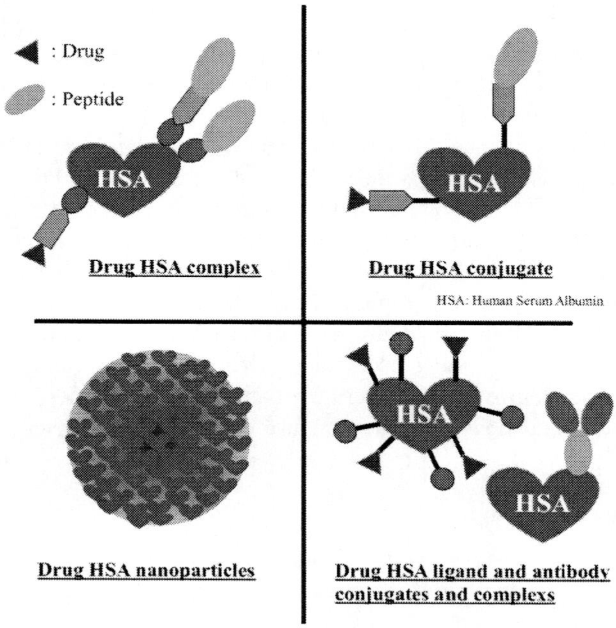

Figure 1. Albumin, a versatile drug carrier for developing drug-, peptide- or antibodybased drugs as conjugates, complexes or nanoparticles.

2. Update on Market Approved Technologies

2.1. Levemir® for the Treatment of Diabetes

Levemir®, an albumin-binding derivative of human insulin, for treating types 1 and 2 diabetes has opened a new field of peptide drug delivery.

The success and mode of action of Levemir® is based on a very simple principle: the C-terminal amino acid threonine in recombinant human insulin is replaced by a lysine moiety, and myristic acid is then covalently bound to its ε-amino group. The affinity of fatty acids such as myristic acid for HSA with at least five binding sites is high with K_d values of ~0.05 μM. After subcutaneous injection, the half-life of Levemir® is extended from 4 to 6 min for native human insulin to 5 to 7 h for Levemir® and for most patients, especially patients with juvenile diabetes (type 1 diabetes), one subcutaneous injection per day is sufficient to normalize blood glucose levels [7].

Diabetes mellitus is caused by a lack of insulin production by the islets of Langerhans in the pancreas (type 1 diabetes) or insulin resistance (type 2 diabetes). In contrast to competing products, Levemir® is the only long-acting insulin that remains soluble both before and after injection and does not form microprecipitates or crystals, unlike insulin glargine and NPH insulin.

The biologically active forms, GLP-1-(7–37) and GLP-1-(7–36)NH_2, are produced after the cleavage of proglucagon in the gut and stimulate the secretion of insulin by pancreatic cells. GLP-1(7–37) has a half-life of 1.5–2 min due to degradation by ubiquitous enzymes. Liraglutide is an albumin-binding derivative of GLP-1 in which this peptide hormone is derivatized with myristic acid at the ε-amino position of lysine introduced at the N-terminal position of glutamic acid in the GLP-1 peptide sequence. In contrast to GLP-1, liraglutide is stable against metabolic degradation due to albumin-binding and has a plasma half-life of 11–15 h after subcutaneous administration making it suitable for once daily administration.

Table 1. Overview of some albumin-based drugs that have reached market approval or are in different stages of clinical development

Product	Drug	Indication	Current status
•ABI-007 (Abraxane®)	Albumin-paclitaxel nanoparticle	•Oncology	Marketed
•^{99m}Tc-Albures	^{99m}Tc –aggregated albumin	•Oncology	Marketed
•^{99m}Tc-Nanocoll	^{99m}Tc –aggregated albumin	•Oncology	Marketed
•Vasovist®	Albumin-binding Gadolinium (III) complex	•Oncology	Marketed
•B-22956/1	Albumin-binding Gadolinium (III) complex	•Oncology	Marketed
•Levenir®	Albumin-binding fatty acid derivative of insulin	•Diabetes	Marketed
•Liraglutide (Victoza®)	Albumin-binding fatty acid derivative of GLP-1	•Diabetes	Marketed
•AT-103 (Ozoralizumab)	Albumin-binding nanobody directed against human TNF-α	•Rheumatology	Phase II
•INNO-206	Albumin binding prodrug of doxorubicin	•Oncology	Phase II
•ABI-008	Albumin-docetaxel nanoparticle	•Oncology	Phase II
•MTX-HSA	Methotrexate albumin conjugate	•Oncology	Phase I/II
•AFL-HSA	Albumin conjugate of aminofluorescein	•Oncology	Phase I/II
•CjC-1134-PC	Albumin conjugate of exendin-4	•Diabetes	Phase I/II
•ABI-009	Albumin-rapamycin nanoparticle	•Oncology	Phase I
•ABI-010	Albumin nanoparticle with a HSP90 inhibitor	•Oncology	Phase I

Table 2. Overview of typical albumin fusion partners and their potential therapeutic applications

Albumin-fusion products	Potential application	Features
·Alb-growth hormone (Albutropin™)	Growth hormone deficiency	·Clinical trial
·Alb-granulocyte colony- stimulating factor (GCSF)	Neutropenia	·Clinical trial
·Alb-interferon (IFN)- α2b (Albinterferon α-2b)	Chronic hepatitis C	·Clinical trial
·Alb-interleukin (IL)-2 (albuleukin)	Oncology	·Clinical trial
·Alb-IL-1 receptor antagonist	Rheumatoid arthritis	·Animal model study
·Alb-glucagon-like pepetide (GLP) 1 (albuglutide)	Diabetes mellitus	·Clinical trial
·Alb-atrial natriuretic pepetide (ANP) ·Alb-B-type natriuretic pepetide (BNP)	Hypertension, Congestive heart failure	·Animal model study
·Alb-tumor necrosis factor (TNF) ligands ·Animal model study (TNF-α, TWEAK, TRAIL)	Oncology	·Animal model study
·Alb-thymosin-α_1	Hepatitis B and C, malignancy	·Animal model study
·Alb-vascular endothelial growth factor (VEGF)165b	Oncology	·Animal model study
·Alb-onconase	Oncology	·Animal model study
·Alb-tissue inhibitor of metalloproteinase (TIMP)-2 ·^{123}I- and ^{68}Ga-labeled cRGD-Alb-TIMP-2	Oncology	·Animal model study
·Alb-factor VIIa	Haematophilia A	·Animal model study
·Alb-coagulation factor IX	Haemophilia B	·Clinical trial

2.2. Abraxane® for Treating Solid Tumors

Abraxane®, an albumin-paclitaxel nanoparticle, is the most advanced drug product of the so-called nab technology developed by American Bioscience in which a lipophilic drug and HSA are passed through a jet under high pressure to form nab-paclitaxel nanoparticles with a mean particle size of 130 nm. The resulting paclitaxel nanoparticle, Abraxane®, is water-soluble and was first approved by the FDA for the treatment of metastatic breast cancer and since then, has been approved in over 40 countries around the world.

Dosed at a higher maximum tolerated dose of 260 mg/m^2 taxol equivalent compared with Taxol®/CremophorEL at 175 mg/m^2 on a 3-weekly schedule, a randomized phase III study in 460 patients with metastatic breast cancer that was relevant for approval by the FDA demonstrated statistically significant higher response rates (33% versus 19%, P = 0.001) longer time to tumor progression (23.0 versus 16.9 weeks, P = 0.006), and increased survival in the subset of patients receiving second-line or greater treatment (65.0 versus 55.7 weeks, P = 0.024) [8]. Abraxane® avoids the use of cremophor, and appears to be generally well tolerated as a 30 min intravenous infusion, and despite the lack of pre-medication with corticosteroids no hypersensitivity reactions have been reported. These encouraging results were also seen in a comparison of Abraxane® with Taxotere® (polysorbate-based docetaxel) in a randomized phase II clinical trial of the first-line treatment of 300 patients with metastatic breast cancer [9] at lower doses on a weekly schedule, either at 100 mg/m^2 or 150 mg/m^2 taxol equivalents compared with Taxotere® at 100 mg/m^2 on a 3-weekly schedule. Progression-free survival was increased by approximately 5 months and both doses of Abraxane® were associated with an improved safety and toxicity profile compared with docetaxel.

American Bioscience has extended the evaluation of Abraxane® to other tumor indications including non-small cell lung carcinoma (NSCLC), pancreatic cancer, melanoma, and head and neck cancer, often in combination with conventional chemotherapy [10, 11].

One of the most interesting aspects of Abraxane® is the detailed insights into its mode of action that has been obtained preclinically as well as clinically. Abraxane®, although stable as a nanoparticle in its galenic formulation, dissolves rapidly after intravenous infusion resulting in the formation of soluble albumin-bound paclitaxel complexes having a size basically comparable to that of endogenous albumin. The albumin-paclitaxel complexes accumulate in the tumor through the well-established enhanced and permeation effect of solid tumors, but a further albumin transport pathway mediated by the 60-kDa glycoprotein gp60, also known as albondin, which is located on the endothelial cell surface, appears to be responsible for tumor uptake and above all for the even tumor distribution of Abraxane® and the subsequent release of paclitaxel [12-14].

Albumin-paclitaxel complexes bind to gp60 with a high affinity in the nanomolar range inducing gp60 clustering and transcytosis in which both gp60-bound and fluid phase albumin are transported through the tumor endothelium subsequently releasing paclitaxel into the subendothelial space (Figure 2).

Upon entering the tumor interstitium, the accumulation of albumin is possibly facilitated by Secreted Protein, Acidic and Rich in Cysteine (SPARC), a 43 kDa secreted matricellular glycoprotein with a high binding affinity to albumin and significant homology to gp60. The over-expression of SPARC as a key modulator of extracellular matrix proliferation, and cell migration is associated with increased tumor invasion, metastasis, and a poor prognosis in multiple tumor types, and albumin-binding and the subsequent uptake by tumor cells has been demonstrated in several *in vitro* and *in vivo* tumor models [15, 16].

Clinical studies strongly suggest that SPARC is an important participant in the mechanism responsible for the action of Abraxane® contributing to its improved clinical efficacy versus Taxol®/CremophorEL. In the above-mentioned phase I/II study in patients with metastatic pancreatic cancer who were treated with the combination of Abraxane® of 125 mg/m^2 paclitaxel equivalents and gemcitabine at a dose of 1000 mg/m^2 a median overall survival of 12.2 months was achieved. A sub-analysis of the overall SPARC levels in 36 of the 44 evaluable patients showed that a higher SPARC expression was correlated with an overall improved survival ($n = 20$; median survival 13.6 months, $P = 0.02$) compared to patients with low SPARC levels ($n = 16$; median survival 8.1 months).

In a retrospective analysis of 16 patients with head and neck cancer, this trend for SPARC-positive patients was confirmed. Based on these observations, the role of SPARC as a predictive biomarker for the clinical response to Abraxane® is currently being investigated in several clinical trials. The gp60 receptor also most likely plays an important role for the homogeneous tumor tissue distribution and deep tumor penetration of albumin-binding bispecific antibodies and nanobodies. In addition to Abraxane®, the nab technology platform has been successfully used to encapsulate other anticancer lipophilic drugs such as docetaxel, the mTOR inhibitor rapamycin, and the heat-shock protein inhibitor Hsp90. The efficacy of these albumin drug nanoparticles as a single agent has been demonstrated in preclinical studies in multiple tumor xenograft models.

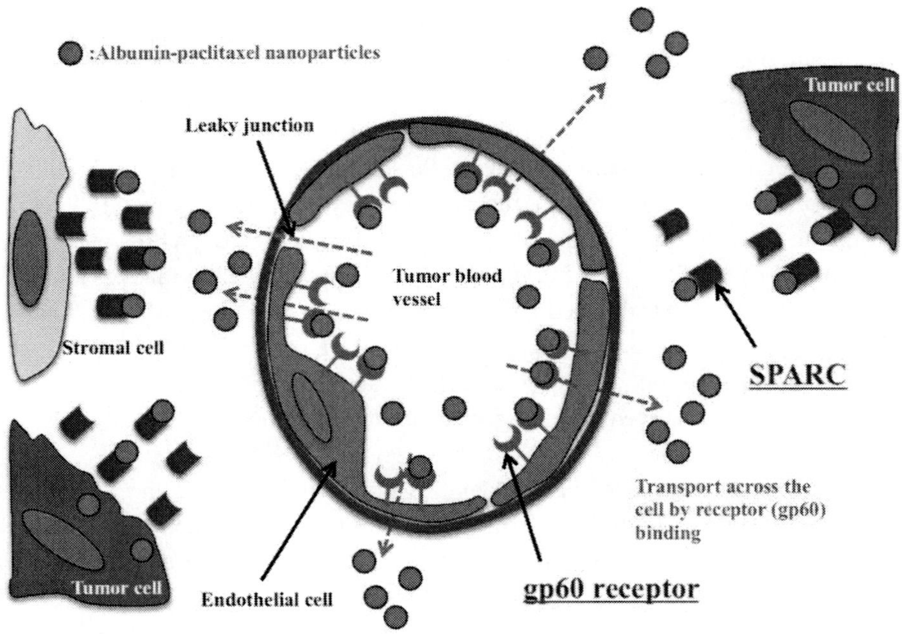

Figure 2. Uptake of albumin-paclitaxel nanoparticles is presumably mediated by the gp60 transcytosis pathway and subsequent binding to SPARC (Secreted Protein, Acidic and Rich in Cysteine) in the tumor extracellular matrix.

3. Update on Albumin-Binding Prodrugs and Drug Albumin Conjugates in Clinical Trials

3.1. INNO-206 for Clinical Trials against Sarcoma and Gastric Cancer

Kratz et al. showed that a maleimide bearing prodrug of doxorubicin was rapidly and selectively bound to the cysteine-34 position endogenous albumin [17]. As a result of further preclinical study, the acid-sensitive (6-maleimidocaproyl)hydrazone derivative of doxorubicin (DOXO-EMCH, renamed INNO-206) emerged as a clinical candidate due to its high plasma stability of its albumin-bound form, the superior efficacy of INNO-206 in eight murine tumor models and a favorable toxicity profile including a significantly reduced cardiotoxicity [18, 19].

Preclinically as well as clinically, the albumin-bound form of INNO-206 has a large AUC, a small volume of distribution and low clearance compared to doxorubicin, its uptake in solid tumors being mediated by the pathophysiology of tumor tissue, as characterized by angiogenesis, hypervasculature, a defective vascular architecture, and impaired lymphatic drainage. The prodrug contains an acid-sensitive hydrazone linker, which allows doxorubicin to be released either extracellularly in the slightly acidic environment often present in tumor tissue or intracellularly in acidic endosomal or lysosomal compartments after the cellular uptake of the albumin conjugate by the tumor cell. In a phase I study, INNO-206 showed a good safety profile at doses up to 260 mg/m^2 doxorubicin equivalents [20]. In a recent preclinical study, a test was conducted as to whether a combination of suboptimal doses of INNO-206 with doxorubicin would improve therapeutic efficacy compared to their antitumor effect when administered at their respective maximum tolerated doses (MTDs) as single agents [21].

3.2. Albuferon® for Treating Hepatitis C

Albuferon® "Albinterferon-α-2b" is a fusion protein of albumin and interferon-α-2b (INFα-2b) that is used in the treatment of hepatitis C infections which affects approximately 180 million people worldwide [22]. INFα-2b has a molecular weight of ~ 19 kDa and, despite its antiviral properties, has a half-life in humans in the range of 2 to 3 h, thus requiring frequent injections. By genetically fusing recombinant INFα-2b to recombinant HSA a fusion protein with a molecular weight of 85.7 kDa is generated that has been developed by Human Genome Sciences in collaboration with Novartis as a long-acting interferon for the treatment of chronic hepatitis C and has been evaluated in several phase III trials most of which are now complete.

Several phase III programs have compared the efficacy, safety and impact on health-related quality of life of Albinterferon-α-2b versus Pegasys® (INFα-2b pegylated with a branched 40 kDa poly(ethylene)glycol) representing the current standard of care in combination with the antiviral nucleoside ribavirin, [23]. Another phase III study was conducted to assess the efficacy/safety of Albinterferon-α-2b in patients with chronic hepatitis C virus genotype 2/3. In this study, 933 patients were randomized to an open-label subcutaneous treatment with Pegasys® (180 μg once a week) or Albinterferon-α-2b (900 or 1200 μg every 2 weeks) for 24 weeks, each administered with oral ribavirin 800 mg/day. As a main conclusion, this study showed that Albinterferon-α-2b was not inferior to Pegasys® in patients with chronic hepatitis C virus genotype 2 or 3 and Albinterferon-α-2b 900 μg administered every 2 weeks provides an alternative efficacious treatment option for such patients. A further recently completed phase III trial study including 1331 patients with chronic hepatitis C virus genotype 1, revealed that Albinterferon-α-2b was not inferior to Pegasys® in these patients and treatment with Albinterferon-α-2b at a dose of 900 μg every 2 weeks showed comparable efficacy with a similar toxicity profile but a higher discontinuation rate [24].

4. The Use of Albumin As a Drug Carrier for Antibodies and Therapeutic Peptides — Preclinical and Clinical Progress

The strategy of binding peptides, e.g., cytokines, various antibody fragments, to albumin is meanwhile being extensively explored for the design of a new generation of albumin-binding antibodies and therapeutically active peptides. Three strategies can be distinguished: (i) bivalent antibodies that bind the desired antigen or peptide derivatives are constructed that additionally contain the heavy or light chains of antibodies as an albumin-binding domain; (ii) the Fc region is substituted by albumin and two antigen-binding regions (Fab regions) are fused to albumin; and (iii) an albumin-binding peptide is bound to the Fab regions, other antibody fragments, or therapeutic peptides that bind to serum albumin with high affinity in the nano- to fentomolar range.

The rationale for these approaches is to exploit the long-half life of albumin in the body, which is approximately 19 days in humans and/or the targeting properties of albumin for inflamed and malignant tissue. In this context, the long half-life of albumin cannot be explained by its molecular-weight (66.5 kDa) alone but, rather, as demonstrated in recent research, albumin binds to the neonatal Fc receptor (FcRn) which also binds IgGs and is responsible for the long-half life of this class of antibodies which is also in the range of 2–3 weeks. FcRn is a cell membrane bound receptor, which is found on endothelial cells, but also on cells of various organs such as the kidneys, liver and intestine. The binding of serum albumin to FcRn was discovered and elucidated by C.L. Anderson and co-workers who showed that a similar to IgG was bound to serum albumin in a pH dependent manner [25, 26].

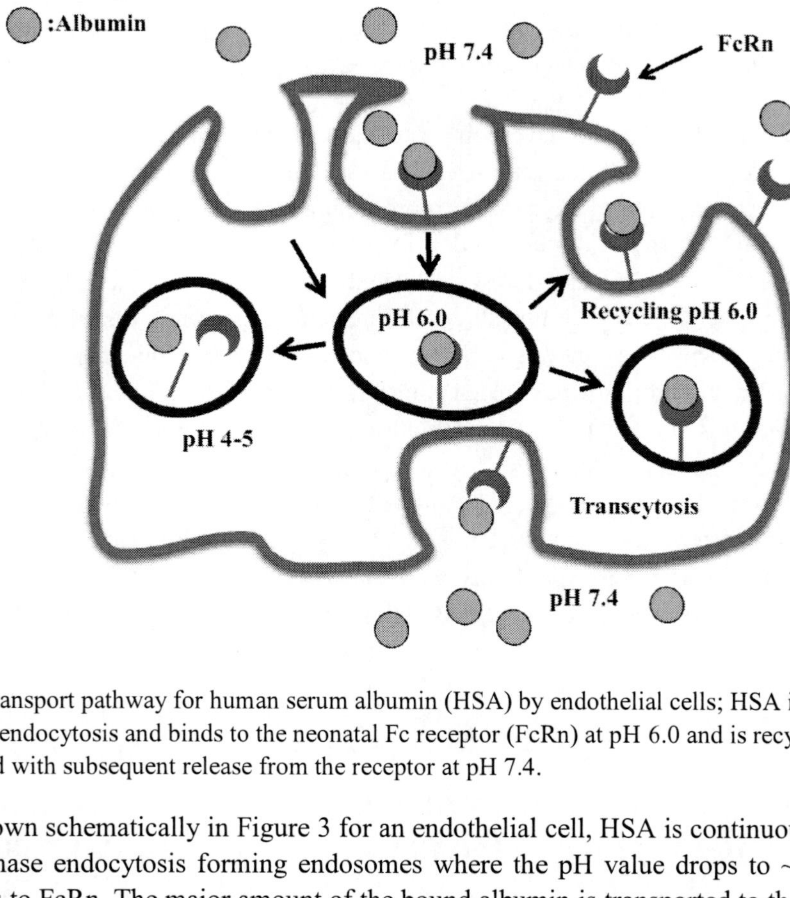

Figure 3. Transport pathway for human serum albumin (HSA) by endothelial cells; HSA is taken up by fluid-phase endocytosis and binds to the neonatal Fc receptor (FcRn) at pH 6.0 and is recycled or transcytosed with subsequent release from the receptor at pH 7.4.

As shown schematically in Figure 3 for an endothelial cell, HSA is continuously taken up by fluid-phase endocytosis forming endosomes where the pH value drops to ~ pH 6.0 and HSA binds to FcRn. The major amount of the bound albumin is transported to the interstitium or recycled where it is released from the receptor at pH 7.4, while a smaller fraction undergoes lysosomal degradation. The elucidation of the pH-dependent binding of serum albumin to FcRn that controls its half-life is an important issue in designing albumin-based therapeutic or diagnostic agents for modulating their pharmacokinetics and drug targeting properties.

4.1. Novel Single Chain Antibodies with High Affinity Albumin-Binding Peptide Domains

Dennis and co-workers at Genentech, Inc. published pioneering work on the attachment of albumin-binding single chain antibodies in which they discovered a peptide comprised of 18 amino acids (Ac-RLIEDICLPRWGCLWEDD-NH_2) with a high binding affinity for HSA (K_d ~ 0.5 μM). A related sequence QRLMEDICLPRWGCLWEDDF comprised of 20 amino acids was then fused with the single chain antibody D3H44 Fab directed against the tissue factor and showed a similar affinity for HSA. When mice or rabbits were injected with the preparation, the plasma half-life was found to be extended to 26-fold for 10.4 h and 37-fold for 34.2 h in rabbits which corresponds to 25–43% of the half-life in these animals. Thus, by attaching a fairly small peptide to a single chain antibody the rapid clearance of the native single chain antibody that generally limits their therapeutic usefulness could be overcome and its pharmacokinetic profile fundamentally altered [27].

As a consequence Dennis and co-workers performed a fascinating experiment comparing the tumor uptake, tumor penetration as well as the biodistribution of Herceptin® (trastuzumab) of a single chain antibody Fab4D5 directed against HER2 and of AB.Fab4D5 that additionally contained the albumin-binding sequence QRLMEDICLPRWGCLWEDDF attached [28].

HER2 (also known as ErbB-2) is the abbreviation for "Human Epidermal growth factor Receptor 2" and is an established tumor marker that is over-expressed in ~25% of breast cancers and a prognostic indicator of long-term survival in tumor patients [29]. The more even tumor tissue distribution of the single chain antibody can be explained by the much smaller size of Fab4D5 (~ 25 kDa). The interstitial fluid pressure of tumors hinders the uptake of macromolecules which is mediated primarily by convection and not diffusion. The albumin-binding antibody AB.Fab4D5, once bound to albumin, has a molecular weight of ~ 90 kDa and although this is less than the 150 kDa for an IgG, it is unlikely that this smaller size alone is the reason for its more efficient and even tumor uptake. Rather, it is more likely that the high-affinity receptor for albumin, albondin (gp60), which is present on the surface of the tumor endothelium, transports AB.Fab4D5 through transcytosis into the inner core of the tumor resulting in an active transport mechanism counterbalancing the interstitial fluid pressure in the inner region of the tumor. As discussed above, the gp60 receptor plays an important role in the tumor tissue distribution of albumin-bound paclitaxel (Abraxane®) and upon entering the tumor interstitium, the accumulation of AB.Fab4D5 is possibly facilitated by SPARC, especially given the fact that this matricellular glycoprotein is abundantly secreted in breast cancer [30].

It is clear that rapid targeting, excellent tumor deposition and retention, coupled with high tumor to blood ratios may make AB.Fab4D5 an exceptional molecule for use in imaging and cancer therapy. Although Herceptin® has improved overall survival in first-line metastatic breast cancer in HER2-positive patients ranging from 5 to 8 months depending on which anticancer agent it was combined with (either doxorubicin, cyclophosphamide, paclitaxel or docetaxel) [31, 32], the use of trastuzumab either in late-stage treatment or in an adjuvant setting has disadvantages that include high costs of up to US$ 100,000 per year, the development of resistance and cardiac dysfunction in 2–7% of cases. Indeed, approximately 10% of eligible patients are unable to tolerate this drug because of pre-existing heart problems

[33]. Thus, the use of albumin-binding single chain antibodies such as AB.Fab4D5 could circumvent these complications due to an improved tumor tissue distribution and altered biodistribution.

4.2. The Albumod™ Technology Which Extends the Half-Life of Biopharmaceuticals through a High-Affinity Albumin-Binding Domain

The Affibody® technology is based on a similar approach of enhancing the the pharmacokinetic profile of therapeutic peptides, but uses high affinity peptide molecules with molecular weights of ~ 5–7 kDa which exhibit an extremely high affinity for several proteins including HSA with binding constants in the picomolar or even fentomolar range [34]. Engineered Affibody® peptides have primarily been evaluated preclinically using HER2-positive tumor models, either diagnostically [35-37] or therapeutically [38, 39].

The Albumod™ technology makes use of the high affinity binding albumin-binding domains (ABD) that extend the half-life of conjugated biopharmaceuticals as the result of their extremely strong binding to albumin. As a consequence, the resulting albumin drug complex essentially acquires pharmacokinetic properties that are similar to endogenous albumin and bound cytokines, single chain antibodies or drugs containing the ABD therefore have a significantly increased plasma half-life.

As an example, Affibody can bind granulocyte colony-stimulating factor (G-CSF), a peptide growth factor that stimulates the bone marrow, to produce granulocytes and stem cells, to their ABD. A recombinant form of G-CSF, Filgrastim (Neupogen®) is used with cancer patients to accelerate recovery from neutropenia after chemotherapy, but due to its relatively low molecular weight of ~ 19 kDa is has an elimination half-life of only ~ 3.5 h. The conjugate of G-CSF and ABD, G-CSF-ABD, forms a stable complex with albumin, with an increased plasma-half-life. In contrast to conventional antibodies, Affibody's Albumod™ technology could have the advantages of facilitating a more economic production over genetically engineered albumin fusion proteins.

5. Albumin As a Carrier of Bioactive Gases (Bio-Gas) Such As Nitric Oxide

Nitric oxide (NO) is a unique, diffusible molecular messenger that plays a central role in mammalian physiology and pathophysiology [40-46]. The effects of NO are pleiotropic, including vascular smooth muscle relaxation [47, 48], inhibition of platelet aggregation [49], and the regulation of immune and neuronal functions [50]. However, under certain circumstances NO can be cytotoxic. For example, high concentrations of NO can inhibit tumor cell growth and induce apoptosis. Recent studies have revealed that NO is associated with, not only apoptosis of cancer cells, but also with the inhibition of cancer progression and metastasis, as well as cancer angiogenesis. It also functions as a modulator for chemo/radio/immuno-therapy. Unfortunately, despite such highly useful properties, the use of NO has been impeded by the fact that its *in vivo* half-life is so extremely short (~0.1 s) that

NO itself cannot be used as a therapeutic agent. Therefore, pharmacologically active compounds that can release NO or lead to its formation in the body have been prepared. For example, organic nitrates and nitrite esters have been used for many years to treat patients with ischemic heart disease. However, there are well-known side-effects and limitations to the use of these NO donors, including potentially adverse hemodynamic effects, drug tolerance, lack of selectivity and limited bioavailability [51, 52]. Thus, it is essential to develop reliable NO donors with better pharmacological and pharmacokinetic parameters.

In our search for a reliable and safe NO donor, we followed a different approach, namely examining the possibility of using a NO-traffic protein. The term NO-traffic protein is meant to be a protein with i) a high efficiency of *S*-nitrosation, ii) a high stability of the *S*-nitroso form in the circulation and iii) a high efficiency of *S*-transnitrosation in cells which require NO. As a candidate, we focused on HSA, because HSA is the most abundant plasma protein (35-50 g/L), and because 'endogenous' *S*-nitrosothiols in human plasma are largely associated with HSA [53, 54].

5.1. Native, *S*-Nitrosated HSA As a NO Carrier

S-nitrosated HSA (SNO-HSA) is significantly more stable than low molecular weight *S*-nitrosothiols. In addition to our group, other researchers have also attempted to produce NO delivery systems using a NO-albumin conjugates. Thus, Marks et al. [55] and Ewing et al. [56] synthesized a macromolecular *S*-nitrosothiol, Poly-SNO-bovine serum albumin (BSA), in which several *S*-nitrosothiols are formed in BSA after reduction of the disulfide bonds contained in the protein. In a porcine coronary angioplasty model, Poly-SNO-BSA was applied locally to a site of vascular injury and showed a high retention at the administration site and reduced platelet attachment and activation. These effects were due to the strong binding of the modified albumins to the injured vessels. In the development of targeted NO delivery systems for intravenous use, the tissue distribution characteristics of the NO-carrier conjugate should be evaluated *in vivo* in order to identify the various obstacles to targeted delivery, such as extensive uptake by mononuclear phagocyte systems and rapid loss by glomerular filtration. Katsumi et al. examined the pharmacokinetic properties of SNO-BSA. The results showed that serum albumin is a promising carrier for controlling the pharmacokinetic properties of NO after intravenous injection, because *S*-nitrosated albumin shows a relatively high retention in the blood circulation after intravenous injection in mice. However, targeted NO delivery after intravenous injection using a macromolecular carrier has not been successfully achieved to date [57].

To achieve targeted NO delivery for SNO-HSA after intravenous injection, it will be necessary to understand the method of *S*-nitrosation, the structure of HSA and its biological fate in detail. Therefore, we recently examined the structure and the biological effects of Mono-SNO-HSA, HSA with one (or less) *S*-nitrosothiol, and Poly-SNO-HSA.

Endogenous *S*-nitrosated HSA (Mono-SNO-HSA) is a high molecular weight NO carrier in human plasma, which has shown to have many beneficial effects in different animal models. In an attempt to construct more efficient SNO-HSA preparations, we prepared a SNO-HSA that contained numerous conjugated SNO groups (Poly-SNO-HSA) by chemical modification with 2-iminothiolane. We compared the properties of such a preparation to those of Mono-SNO-HSA using C26 and HepG2 cells [58]. The findings showed that the cellular uptake of NO from Mono-SNO-HSA partly takes place via a low molecular weight thiol, which results in the development of cytoprotective effects by the induction of heme oxygenase-1. By contrast, the transfer of NO from Poly-SNO-HSA into the cells was faster and more pronounced. The influx mainly occurs via the action of a cell-surface protein disulfide isomerase (also called endoplasmic reticulum protein 5). Instead of cytoprotection, the considerable NO inflow resulted in apoptotic cell death caused by the induction of reactive oxygen species (ROS) production, caspase-3 activation and other means. Thus, increasing the number of SNO groups on HSA does not simply intensify the cellular responses to NO but can also result in very different effects (Figure 4). Thus, Mono-SNO-HSA could function as a cytoprotective agent, whereas Poly-SNO-HSA could have the potential to function as an antitumor agent. Here, we wish to discuss the possibility of using Poly-SNO-HSA as a safe and effective multifunctional antitumor agent in biological systems.

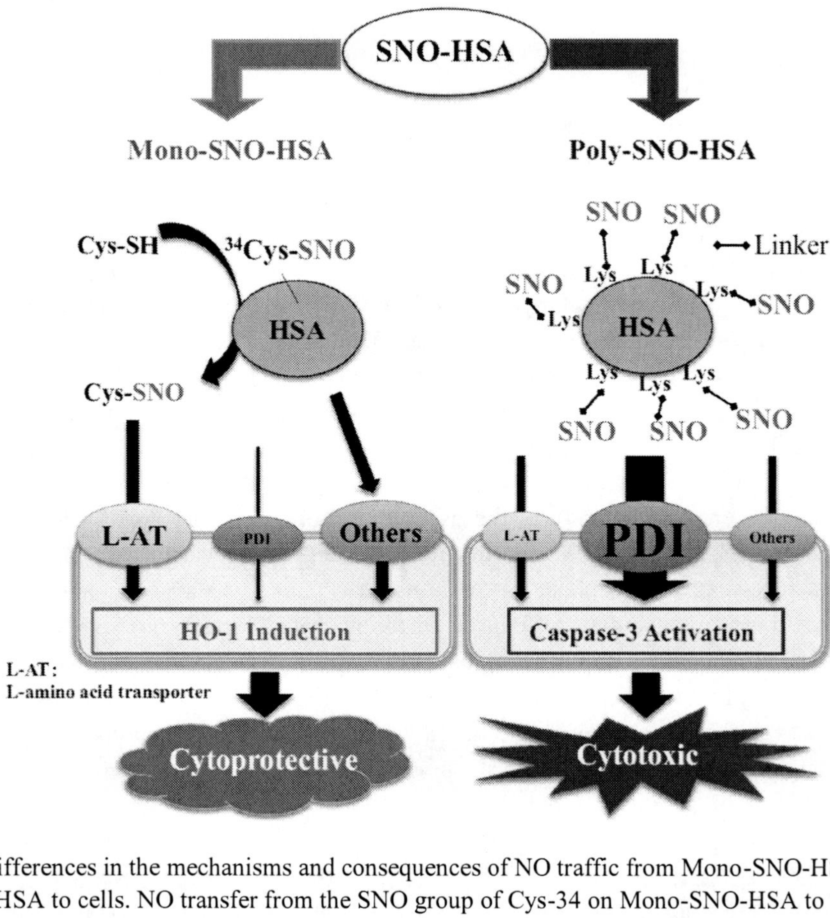

Figure 4. Differences in the mechanisms and consequences of NO traffic from Mono-SNO-HSA and Poly-SNO-HSA to cells. NO transfer from the SNO group of Cys-34 on Mono-SNO-HSA to the cell is partly mediated by the L-amino acid transporter (L-AT) via *S*-transnitrosation to free the low molecular weight thiol. By contrast, NO transfer from Poly-SNO-HSA is mainly mediated by cell-surface protein disulfide isomerase (PDI) without *S*-transnitrosation to free low molecular weight thiol. The relatively slow transfer of NO from Mono-SNO-HSA avoids the presence of high intracellular NO concentrations and leads to cytoprotective activity through heme oxygenase-1 (HO-1) induction. On the other hand, the NO influx from Poly-SNO-HSA is very rapid and pronounced and leads to cell death caused by apoptosis.

5.2. The Direct Antitumor Effect of Poly-SNO-HSA via Apoptosis

Poly-SNO-HSA has been investigated as a NO donor because of potential therapeutic applications, but little is known concerning the effects of Poly-SNO-HSA on cancer. Therefore, our research group produced Poly-SNO-HSA using 2-iminothiolane as the chemical linker and observed that this HSA form is capable of inducing apoptosis in cancer cells [58]. The apoptosis occurs via activation of the intrinsic apoptosis pathway in, e.g., murine colon 26 carcinoma cells and in the rat tumor cell line LY-80 (a variant of Yoshida sarcoma) both *in vivo* and *in vitro*.

The process was studied in some detail. The findings indicate that mitochondria appear to play a pivotal role in the regulation of this process in mammals, because it is believed that loss of mitochondrial membrane potential is an essential element of apoptosis. To evaluate the effects of Poly-SNO-HSA on mitochondrial function and membrane potential, LY-80 cells were loaded with a mitochondrial-selective fluorescent cation (rhodamine 123), and we found that a Poly-SNO-HSA treatment decreased rhodamine fluorescent intensity in a dose-dependent manner [59, 60]. In addition, Caspase-3 is a cell-death protease that is involved in the downstream execution phase of apoptosis, during which cells undergo morphological changes, such as DNA fragmentation, chromatin condensation, and the formation of apoptotic bodies. Compared to HSA, LY-80 cells treated with 25, 50 or 100 μM Poly-SNO-HSA showed relative increases in caspase-3 activity of 21-, 34- and 42-fold, respectively. As an interesting observation supporting this result, the uptake of NO from Poly-SNO-HSA to HepG2 cells was also much higher than that for Mono-SNO-HSA. To further confirm that Poly-SNO-HSA induced apoptosis in LY-80 cells, DNA fragmentation, which is a morphological change characteristic of the execution phase of apoptosis, was examined. DNA fragmentation was observed in LY-80 cells after a 4 h of incubation with 100 μM Poly-SNO-HSA and became stabilized after 8 h.

To investigate the anti-tumor effects of Poly-SNO-HSA *in vivo*, murine colon 26 carcinoma tumor-bearing rats were given an intravenous injection of Poly-SNO-HSA. Interestingly, tumor growth in animals that received an intravenous injection of Poly-SNO-HSA was one-third that observed for the HSA-treated animals. This observation suggests that Poly-SNO-HSA has anti-tumor effects *in vivo*, presumably due to its ability to induce apoptosis. We observed similar results, when we used mice bearing other cell types such as SW480 cells.

5.3. Poly-SNO-HSA Can Overcome MDR

Cancer treatment remains one of the most important clinical challenges to biomedical science. One difficulty in treating various cancers is the development of multidrug resistance (MDR) by cancer cells. Various approaches have been tested to overcome MDR, including the use of agents that inhibit P-glycoprotein (P-gp) directly or indirectly through altering the cell membrane and using targeted drug delivery. Most of the approaches have shown some success in small animal models but clinical applications remain limited. However, NO is one of the compounds that have shown promises in treating cancer.

In addition to its apoptosis effects, the antitumor effect of Poly-SNO-HSA could be due to its ability to overcome MDR. In fact, a significant attenuation in resistance to dx in K562/dx cells was found as the concentration (0.5-10 μM) of Poly-SNO-HSA increased. The inhibitory effects of Poly-SNO-HSA and dx on the growth of K562/dx cells were evaluated for synergistic action using isobologram analysis. These results indicate that Poly-SNO-HSA has the following two antitumor effects: it inhibits cell growth and attenuates resistance to dx. [61]

To investigate the effect of Poly-SNO-HSA on chemo-resistance *in vivo*, we prepared K562 and K562/dx tumor-bearing mice by injecting K562 or K562/dx cells into the left hind flank of female BALB/cAJcl-nu/nu mice. After the tumors had grown to 150-200 mm^3, the mice received either intraperitoneal or intravenous injections of either saline, dx and/or Poly-SNO-HSA biweekly. The dx treatment resulted in a significant decrease in the time-dependent tumor growth in K562 tumor-bearing mice. In contrast, in the K562/dx mice dx had no significant effect on tumor growth. A slightly more pronounced, but significant, effect was observed, when the K562/dx mice were treated with Poly-SNO-HSA. However, the use of a combination of dx and Poly-SNO-HSA, resulted in a decrease in tumor volume to one-third compared with treatment with them alone. This observation indicates that Poly-SNO-HSA overcomes dx resistance *in vivo*. Our findings suggest that Poly-SNO-HSA has the potential for use as a safe and strong, multifunctional antitumor agent. This is especially true if the Poly-SNO-HSA could be formulated into suitable carriers targeted to tumors. One of the limitations of the current nanovehicles for targeted drug delivery is that the vehicles also accumulate in non-target organs, causing serious side effects. Since NO is rather benign unless the concentration is too high, it presents a unique opportunity for treating tumors without significant side effects. The current approach of nanotechnology-based drug delivery to tumors can benefit significantly through an understanding of the underlying mechanisms of drug actions, e.g., the therapeutic mechanisms of NO [62].

5.4. Poly-SNO-HSA As an Autophagy Inhibitor

The hypothesis that tumor progression can be curbed via the use of antiangiogenic agents targeting abnormal tumor vessels is supported by preclinical evidence and by clinical trials [63]. However, these initial successes were tempered by the failure of angiogenesis inhibitors to produce long lasting clinical responses. For example, in clinical trials of the vascular endothelial growth factor- (VEGF-) neutralizing antibody bevacizumab in glioblastomas, 40% to 60% of the tumors progressed after an initially successful treatment [64], consistent with the development of resistance to antiangiogenic therapy, a state showing a poor prognosis and poor response to available treatments developed [65]. The molecular basis of acquired resistance to antiangiogenic treatments causing this lack of sustained responses remains unclear. However, Hu et al. demonstrated that the devascularization caused by antiangiogenic therapy increases tumor hypoxia and that this hypoxia mediates resistance to antiangiogenic therapy [66]. Furthermore, recent reports suggest that hypoxia activates autophagy, a lysosomal degradation pathway which may promote the survival of tumor cells [67, 68]. In fact, the devascularization caused by antiangiogenic therapy increases tumor hypoxia, and this hypoxia mediates the resistance to antiangiogenic therapy. The mechanisms by which hypoxia induces autophagy need clarification, but the finding that BNIP3, a HIF-1α downstream target gene, is essential to hypoxia-induced autophagy suggests one possible mechanism [69, 70]. Therefore, autophagy inhibitors may help prevent resistance to antiangiogenic therapy. In our preliminary studies, we found that Poly-SNO-HSA possesses anti-autophagy activity via the inhibition of HIF-1α in tumor cells *in vivo* [71] (Figure 5). We hope that future studies will clarify the issue of whether Poly-SNO-HSA also can be used as a specific and potent autophagy inhibitor.

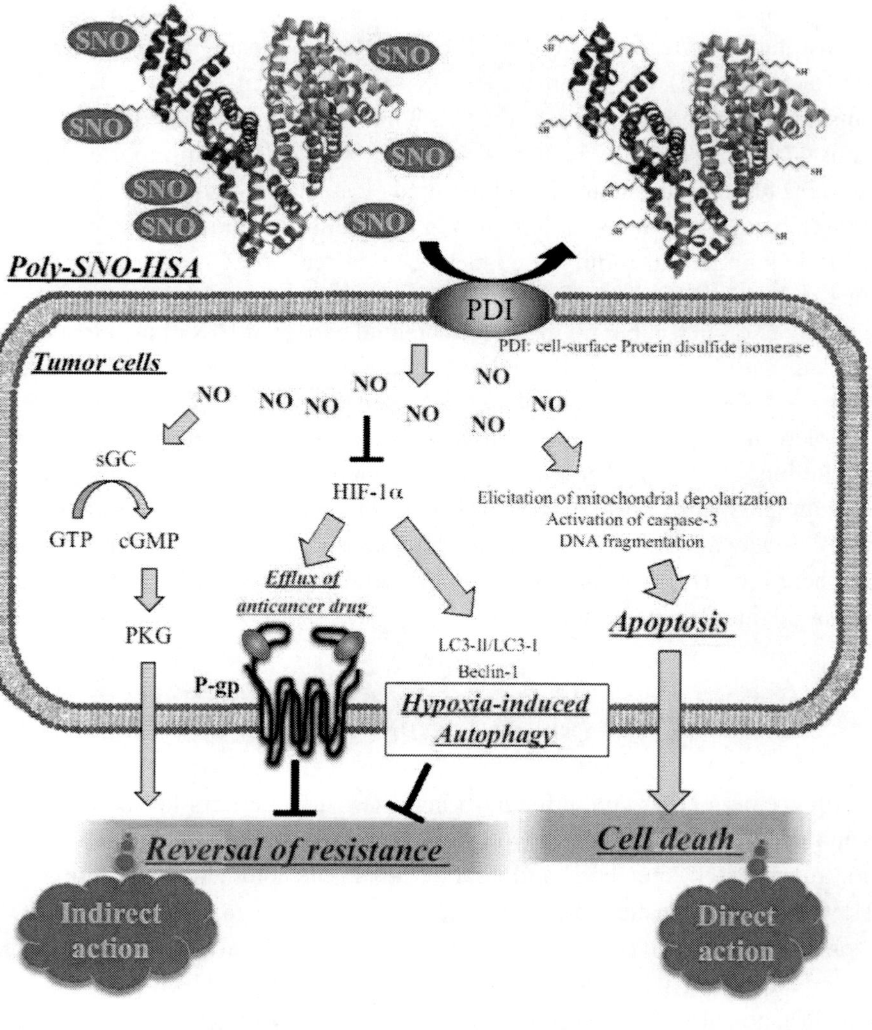

Figure 5. Mechanisms of Poly-SNO-HSA as a safe and strong multiple antitumor agent. The rapid and pronounced transfer of NO from Poly-SNO-HSA into the cell mainly takes place via cell-surface protein disulfide isomerase (PDI). NO also reverts dx resistance partly by activating a cGMP dependent pathway. Finally, NO reverts drug resistance by decreasing the efflux of the drug. The latter effect is brought about by decreasing the expression of hypoxia-inducible factor-1α (HIF-1□α) and P-glycoprotein (P-gp). In addition, Poly-SNO-HSA inhibits hypoxia-induced autophagy via down-regulation of the cell signaling factors LC3-II/LC3-I and Beclin-1.

5.5. HSA As a H_2S Carrier

H_2S, once produced in mammalian cells, can directly exert its biological effects via interaction with different signaling molecules, as occurs with the production of NO from NO synthases or CO from heme oxygenases. H_2S can also have its sulfur stored first and released later in response to a physiological signal. The two main forms of sulfur stores in mammalian cells are acid-labile sulfur and sulfane sulfur [72-74]. Among them, Sulfane sulfur carries no charge. It is always attached to its carrier proteins via covalent bonding between the S_0 atom with its six valence electrons and other sulfur atoms. The existing forms of sulfane sulfur include thiosulfate ($S_2O_3^{2-}$) persulfides (R-S-SH) thiosulfonate (R-S-SO_2-R) polysulfides (RSnR,R is alkyl or aryl), polythionates (SnO_6^{2-}) elemental sulfur (S_8) and disulfides which have an unsaturated carbon adjacent to the C-S bond. A common S_0-binding domain has been identified in 47 human genes [75], including HSA. Westley's group reported that, out of 12 diverse proteins tested, only serum albumin bound elemental sulphur [76].

Treating albumin with a disulfide reducing aggent such as thioredoxin revealed that H_2S is stored in albumin in the form of sulfanes. The current work is consistent with previous suggestions that various 'pools' of biologically relevant sulfur/sulfide species are present in the circulation [77]. Thus, albumin may be a potential carrier of H_2S in the circulation.

We are currently conducting studies regarding HSA persulfide and polysulfide in human serum samples. The results show that HSA persulfide and polysulfide in the human circulation were decreased by several diseases such as chronic renal failure (unpublished data). In addition, HSA persulfide and polysulfide possess very strong antioxidant activity compared with HSA (unpublished data). These data indicate that endogenous HSA persulfide and polysulfide play an important role in the homeostatic mechanism against oxidative stress diseases. Therefore, HSA persulfide and polysulfide may act as a superior antioxidant optimized for the human circulation.

CONCLUSION AND PERSPECTIVES

Albumin is emerging as one of the most important drug carriers for therapeutically active peptides and drugs as well as antibody design, especially for the treatment and diagnosis of malignant, inflammatory metabolic and viral diseases. The commercial success of long-acting insulin (Levemir®) for treating diabetes by simply attaching myristic acid as the albumin-binding moiety has paved the way for the development of a variety of albumin-binding strategies for improving the pharmacokinetic profile and targeting properties of therapeutic active peptides and antibody fragments. These include (i) the attachment of 2–5 kDa peptides or light or heavy chains of antibodies that have a high affinity for albumin in the nano-, pico- or even fentomolar range to a therapeutic/diagnostic peptide or a single chain antibody, (ii) a trivalent nanobody consisting of an anti-HSA camelid antibody (the VHH chain of a single chain antibody with a MW of 12–15 kDa) to two second camelid antibodies directed against the desired antigen, and (iii) an albumin fusion protein produced by genetic engineering which allows antibodies to be produced in which the Fc part of an IgG is replaced by albumin.

Besides the Enhanced permeability and retention (EPR) effect, the gp60 receptor is also responsible for the uptake of the albumin-paclitaxel nanoparticle Abraxane® by tumors. Subsequently, another albumin-binding protein SPARC binds albumin-bound drugs in the tumor extracellular matrix and facilitates a more even tumor tissue distribution. Water-soluble Abraxane® is approved for treating metastatic breast cancer where it is superior to conventional Taxol®/cremophorEL and has been shown to be encouraging in the treatment of pancreatic and head and neck cancer.

Considering the commercial success of products that use albumin as a drug carrier and the ongoing clinical trials as well as due to the many diverse technologies for improving the pharmacokinetic profile and drug targeting of therapeutic and diagnostic peptides, antibody fragments, as well as low-molecular weight drugs that include peptides, synthetic and natural products and even simple molecules such as nitric oxide, albumin is attracting the interest of biotechnology companies as well as of large pharmaceutical companies, and it is likely that the ongoing pipeline of development will move further albumin-based drugs into the clinical setting.

REFERENCES

[1] Langer, K; Balthasar, S; Vogel, V; Dinauer, N; von Briesen, H; Schubert, D. Optimization of the preparation process for human serum albumin (HSA) nanoparticles. *Int J Pharm.*, 257, (2003),169-180.

[2] Rubino, OP; Kowalsky, R; Swarbrick, J. Albumin microspheres as a drug delivery system: relation among turbidity ratio, degree of cross-linking, and drug release. *Pharm Res.*, 10, (1993), 1059-1065.

[3] Kommareddy, S; Amiji, M. Preparation and evaluation of thiol-modified gelatin nanoparticles for intracellular DNA delivery in response to glutathione. *Bioconjug Chem.*, 16, (2005), 1423-1432.

[4] Azarmi, S; Tao, X; Chen, H; Wang, Z; Finlay, WH; Löbenberg, R; Roa, WH. Formulation and cytotoxicity of doxorubicin nanoparticles carried by dry powder aerosol particles. *Int J Pharm.*, 319, (2006), 155-161.

[5] Rahimnejad, M; Jahanshahi, M; Najafpour, GD. Production of biological nanoparticles from bovine serum albumin for drug delivery. *Afr. J. Biotechnol.*, 5, (2006), 1918–1923.

[6] Elsadek, B; Kratz, F. Impact of albumin on drug delivery--new applications on the horizon. *J Control Release.*, 157, (2012), 4-28.

[7] Morales, J. Defining the role of insulin detemir in Basal insulin therapy, *Drugs*, 67, (2007), 2557–2584.

[8] Gradishar, WJ; Tjulandin, S; Davidson, N; Shaw, H; Desai, N; Bhar, P; Hawkins, M; O'Shaughnessy, J. Phase III trial of nanoparticle albumin-bound paclitaxel compared with polyethylated castor oil-based paclitaxel in women with breast cancer, *J. Clin. Oncol.*, 23, (2005), 7794–7803.

[9] Gradishar, WJ; Krasnojon, D; Cheporov, S; Makhson, AN; Manikhas, GM; Clawson, A; Bhar, P. Significantly longer progression-free survival with nabpaclitaxel compared

with docetaxel as first-line therapy for metastatic breast cancer, *J. Clin. Oncol.*, 27, (2009), 3611–3619.

[10] Hawkins, MJ; Soon-Shiong, P; Desai, N. Protein nanoparticles as drug carriers in clinical medicine, *Adv. Drug Deliv. Rev.*, 60, (2008), 876–885.

[11] Desai, N; Trieu, V; Damascelli, B; Soon-Shiong, P. SPARC expression correlates with tumor response to albumin-bound paclitaxel in head and neck cancer patients, *Transl. Oncol.*, 2(2), (2009), 59–64.

[12] Schnitzer, JE. gp60 is an albumin-binding glycoprotein expressed by continuous endothelium involved in albumin transcytosis, *Am. J. Physiol.*, 262(1 Pt 2), (1992), H246–H254.

[13] Tiruppathi, C; Song, W; Bergenfeldt, M; Sass, P; Malik, AB. Gp60 activation mediates albumin transcytosis in endothelial cells by tyrosine kinase-dependent pathway, *J. Biol. Chem.*, 272(41), (1997), 25968–25975.

[14] Wang, Z; Tiruppathi, C; Minshall, RD; Malik, AB. Size and dynamics of caveolae studied using nanoparticles in living endothelial cells, *ACS Nano*, 3(12), (2009), 4110–4116.

[15] Sage, H; Johnson, C; Bornstein, P. Characterization of a novel serum albuminbinding glycoprotein secreted by endothelial cells in culture, *J. Biol. Chem.*, 259(6), (1984), 3993–4007.

[16] Podhajcer, OL; Benedetti, LG; Girotti, MR; Prada, F; Salvatierra, E; Llera, AS. The role of the matricellular protein SPARC in the dynamic interaction between the tumor and the host, *Cancer Metastasis Rev.*, 27(4), (2008), 691–705.

[17] Kratz, F; Muller-Driver, R; Hofmann, I; Drevs, J; Unger, C. A novel macromolecular prodrug concept exploiting endogenous serum albumin as a drug carrier for cancer chemotherapy, *J. Med. Chem.*, 43(7), (2000), 1253–1256.

[18] Graeser, R; Esser, N; Unger, H; Fichtner, I; Zhu, A; Unger, C; Kratz, F. INNO-206, the (6-maleimidocaproyl hydrazone derivative of doxorubicin), shows superior antitumor efficacy compared to doxorubicin in different tumor xenograft models and in an orthotopic pancreas carcinoma model, Invest. *New Drugs*, 28(1), (2009), 14–19.

[19] Kratz, F., Warnecke, A., Scheuermann, K., Stockmar, C., Schwab, J., Lazar, P., Druckes, P., Esser, N., Drevs, J., Rognan, D., Bissantz, C., Hinderling, C., Folkers, G; Fichtner, I; Unger, C. Probing the cysteine-34 position of endogenous serum albumin with thiol-binding doxorubicin derivatives. Improved efficacy of an acid-sensitive doxorubicin derivative with specific albumin-binding properties compared to that of the parent compound, *J. Med. Chem.*, 45(25), (2002), 5523–5533.

[20] Unger, C; Haring, B; Medinger, M; Drevs, J; Steinbild, S; Kratz, F; Mross, K. Phase I and pharmacokinetic study of the (6-maleimidocaproyl)hydrazone derivative of doxorubicin, *Clin. Cancer Res.*, 13(16), (2007), 4858–4866.

[21] Kratz, F; Fichtner, I; Graeser, R. Combination therapy with the albumin-binding prodrug of doxorubicin (INNO-206) and doxorubicin achieves complete remissions and improves tolerability in an ovarian A2780 xenograft model, Invest. *New Drugs*, (2011)., 2012, 30(4), 1743-9.

[22] Flisiak, R; Flisiak, I. Albinterferon-alpha 2b: a new treatment option for hepatitis C, *Expert Opin. Biol. Ther.*, 10(10), (2010), 1509–1515.

[23] Nelson, DR; Benhamou, Y; Chuang, WL; Lawitz, EJ; Rodriguez-Torres, M; Flisiak, R; Rasenack, JW; Kryczka, W; Lee, CM; Bain, VG; Pianko, S; Patel, K; Cronin, PW;

Pulkstenis, E; Subramanian, GM; McHutchison, JG. Albinterferon Alfa-2b was not inferior to pegylated interferon-alpha in a randomized trial of patients with chronic hepatitis C virus genotype 2 or 3, *Gastroenterology*, 139(4), (2010), 1267–1276.

[24] Zeuzem, S; Sulkowski, MS; Lawitz, EJ; Rustgi, VK; Rodriguez-Torres, M; Bacon, BR; Grigorescu, M; Tice, AD; Lurie, Y; Cianciara, J; Muir, AJ; Cronin, PW; Pulkstenis, E; Subramanian, GM; McHutchison, JG. Albinterferon Alfa-2b was not inferior to pegylated interferon-alpha in a randomized trial of patients with chronic hepatitis C virus genotype 1, *Gastroenterology*, 139(4), (2010), 1257–1266.

[25] Chaudhury, C; Mehnaz, S; Robinson, JM; Hayton, WL; Pearl, DK; Roopenian, DC; Anderson, CL. The major histocompatibility complex-related Fc receptor for IgG (FcRn) binds albumin and prolongs its lifespan, *J. Exp. Med.*, 197(3), (2003), 315–322.

[26] Chaudhury, C; Brooks, CL; Carter, DC; Robinson, JM; Anderson, CL. Albumin binding to FcRn: distinct from the FcRn–IgG interaction, *Biochemistry*, 45(15), (2006), 4983–4990.

[27] Dennis, MS; Zhang, M; Meng, YG; Kadkhodayan, M; Kirchhofer, D; Combs, D; Damico, LA. Albumin binding as a general strategy for improving the pharmacokinetics of proteins, *J. Biol. Chem.*, 277(38), (2002), 35035–35043.

[28] Dennis, MS; Jin, H; Dugger, D; Yang, R; McFarland, L; Ogasawara, A; Williams, S; Cole, MJ; Ross, S; Schwall, R. Imaging tumors with an albuminbinding Fab, a novel tumor-targeting agent, *Cancer Res.*, 67(1), (2007), 254–261.

[29] Slamon, DJ; Leyland-Jones, B; Shak, S; Fuchs, H; Paton, V; Bajamonde, A; Fleming, T; Eiermann, W; Wolter, J; Pegram, M; Baselga, J; Norton, L. Use of chemotherapy plus a monoclonal antibody against HER2 for metastatic breast cancer that overexpresses HER2, *N. Engl. J. Med.*, 344(11), (2001), 783–792.

[30] Podhajcer, OL; Benedetti, LG; Girotti, MR; Prada, F; Salvatierra, E; Llera, AS. The role of the matricellular protein SPARC in the dynamic interaction between the tumor and the host, *Cancer Metastasis Rev.*, 27(4), (2008), 691–705.

[31] Marty, M; Cognetti, F; Maraninchi, D; Snyder, R; Mauriac, L; Tubiana-Hulin, M; Chan, S; Grimes, D; Anton, A; Lluch, A; Kennedy, J; O'Byrne, K; Conte, P; Green, M; Ward, C; Mayne, K; Extra, JM. Randomized phase II trial of the efficacy and safety of trastuzumab combined with docetaxel in patients with human epidermal growth factor receptor 2-positive metastatic breast cancer administered as firstline treatment: theM77001 study group, *J. Clin. Oncol.*, 23(19), (2005), 4265–4274.

[32] Dent, S; Verma, S; Latreille, J; Rayson, D; Clemons, M; Mackey, J; Verma, S; Lemieux, J; Provencher, L; Chia, S; Wang, B; Pritchard, K. The role of her2-targeted therapies in women with her2-overexpressing metastatic breast cancer, *Curr. Oncol.*, 16(4), (2009), 25–35.

[33] Seidman, A; Hudis, C; Pierri, MK; Shak, S; Paton, V; Ashby, M; Murphy, M; Stewart, SJ; Keefe, D. Cardiac dysfunction in the trastuzumab clinical trials experience, *J. Clin. Oncol.*, 20(5), (2002), 1215–1221.

[34] Jonsson, A; Dogan, J; Herne, N; Abrahmsen, L; Nygren, PA. Engineering of a femtomolar affinity binding protein to human serum albumin, *Protein Eng. Des. Sel.*, 21(8), (2008), 515–527.

[35] Lee, SB; Hassan, M; Fisher, R; Chertov, O; Chernomordik, V; Kramer-Marek, G; Gandjbakhche, A; Capala, J. Affibody molecules for in vivo characterization of HER2-

positive tumors by near-infrared imaging, *Clin. Cancer Res.*, 14(12), (2008), 3840–3849.

[36] Chernomordik, V; Hassan, M; Lee, SB; Zielinski, R; Gandjbakhche, A; Capala, J. Quantitative analysis of Her2 receptor expression in vivo by near-infrared optical imaging, *Mol. Imaging*, 9, (4), (2010), 192–200.

[37] Hoppmann, S; Miao, Z; Liu, S; Liu, H; Ren, G; Bao, A; Cheng, Z. Radiolabeled affibody- albumin bioconjugates for HER2-positive cancer targeting, *Bioconjug. Chem.*, 22(3), (2011), 413–421.

[38] Orlova, A; Tran, TA; Ekblad, T; Karlstrom, AE; Tolmachev, V. (186)Re-maSGS-Z (HER2:342) a potential Affibody conjugate for systemic therapy of HER2- expressing tumours, *Eur. J. Nucl. Med. Mol. Imaging*, 37(2), (2010), 260–269.

[39] Tolmachev, V; Orlova, A; Pehrson, R; Galli, J; Baastrup, B; Andersson, K; Sandstrom, M; Rosik, D; Carlsson, J; Lundqvist, H; Wennborg, A; Nilsson, FY. Radionuclide therapy of HER2-positive microxenografts using a 177Lulabeled HER2-specific Affibody molecule, *Cancer Res.*, 67(6), (2007), 2773–2782.

[40] Moncada, S; Palmer, R; Higgs, E. Nitric oxide: physiology, pathophysiology, and pharmacology. *Pharmacol Rev*, 43, (1991), 109-142.

[41] Mizutani, T; Layon, A. Clinical applications of nitric oxide. *Chest*, *110*, (1996), 506-524.

[42] Stamler, J; Singel, D; Loscalzo, J. *Biochemistry* of nitric oxide and its redox-activated forms. *Science*, 258, (1992), 1898-1902.

[43] Nathan, C. Nitric oxide as a secretory product of mammalian cells. *FASEB J*, 6, (1992), 3051-3064.

[44] Lancaster, JJ. Simulation of the diffusion and reaction of endogenously produced nitric oxide. *Proc Natl Acad Sci U S A*, 91, (1994), 8137-8141.

[45] Gaston, B; Reilly, J; Drazen, J; Fackler, J; Ramdev, P; Arnelle, D; Mullins, M; Sugarbaker, D; Chee, C; Singel, D. Endogenous nitrogen oxides and bronchodilator S-nitrosothiols in human airways. *Proc Natl Acad Sci U S A*, 90, (1993), 10957-10961.

[46] Clancy, R; Levartovsky, D; Leszczynska-Piziak, J; Yegudin, J; Abramson, S. Nitric oxide reacts with intracellular glutathione and activates the hexose monophosphate shunt in human neutrophils: evidence for S-nitrosoglutathione as a bioactive intermediary. *Proc Natl Acad Sci U S A*, 91, (1994), 3680-3684.

[47] Furchgott, R; Zawadzki, J. The obligatory role of endothelial cells in the relaxation of arterial smooth muscle by acetylcholine. *Nature*, 288, (1980), 373-376.

[48] Palmer, R; Ferrige, A; Moncada, S. Nitric oxide release accounts for the biological activity of endothelium-derived relaxing factor. *Nature*, 327, (1987), 524-526.

[49] Radomski, M; Palmer, R; Moncada, S. The anti-aggregating properties of vascular endothelium: interactions between prostacyclin and nitric oxide. *Br J Pharmacol*, 92, (1987), 639-646.

[50] Dawson, T; Snyder, S. Gases as biological messengers: nitric oxide and carbon monoxide in the brain. *J Neurosci*, 14, (1994), 5147-5159.

[51] Ignarro, L; Napoli, C; Loscalzo, J. Nitric oxide donors and cardiovascular agents modulating the bioactivity of nitric oxide: an overview. *Circ Res*, 90, (2002), 21-28.

[52] Fung, H. Biochemical mechanism of nitroglycerin action and tolerance: is this old mystery solved? *Annu Rev Pharmacol Toxicol*, 44, (2004), 67-85.

[53] Stamler, J; Jaraki, O; Osborne, J; Simon, D; Keaney, J; Vita, J; Singel, D; Valeri, C; Loscalzo, J. Nitric oxide circulates in mammalian plasma primarily as an S-nitroso adduct of serum albumin. *Proc Natl Acad Sci U S A*, 89, (1992), 7674-7677.

[54] Marley, R; Feelisch, M; Holt, S; Moore, K. A chemiluminescense-based assay for S-nitrosoalbumin and other plasma *S*-nitrosothiols. *Free Radic Res*, 32, (2000), 1-9.

[55] Marks, D; Vita, J; Folts, J; Keaney, JJ; Welch, G; Loscalzo, J. Inhibition of neointimal proliferation in rabbits after vascular injury by a single treatment with a protein adduct of nitric oxide. *J Clin Invest*, 96, (1995), 2630-2638.

[56] Ewing, J; Young, D; Janero, D; Garvey, D; Grinnell, T. Nitrosylated bovine serum albumin derivatives as pharmacologically active nitric oxide congeners. *J Pharmacol Exp Ther*, 283, (1997), 947-954.

[57] Katsumi, H; Nishikawa, M; Yamashita, F; Hashida, M. Development of polyethylene glycol-conjugated poly-S-nitrosated serum albumin, a novel S-Nitrosothiol for prolonged delivery of nitric oxide in the blood circulation in vivo. *J Pharmacol Exp Ther*, 314, (2005), 1117-1124.

[58] Ishima, Y; Yoshida, F; Kragh-Hansen, U; Watanabe, K; Katayama, N; Nakajou, K; Akaike, T; Kai, T; Maruyama, T; Otagiri, M. Cellular uptake mechanisms and responses to NO transferred from mono- and poly-*S*-nitrosated human serum albumin. *Free Radic Res*, 45, (2011), 1196-1206.

[59] Katayama, N; Nakajou, K; Komori, H; Uchida, K; Yokoe, J; Yasui, N; Yamamoto, H; Kai, T; Sato, M; Nakagawa, T; Takeya, M; Maruyama, T; Otagiri, M. Design and evaluation of S-nitrosylated human serum albumin as a novel anticancer drug. *J Pharmacol Exp Ther*, 325, (2008), 69-76.

[60] Katayama, N; Nakajou, K; Ishima, Y; Ikuta, S; Yokoe, J; Yoshida, F; Suenaga, A; Maruyama, T; Kai, T; Otagiri, M. Nitrosylated human serum albumin (SNO-HSA) induces apoptosis in tumor cells. *Nitric Oxide*, 22, (2010), 259-265.

[61] Ishima, Y; Hara, M; Kragh-Hansen, U; Inoue, A; Suenaga, A; Kai, T; Watanabe, H; Otagiri, M; Maruyama, T. Elucidation of the therapeutic enhancer mechanism of poly-S-nitrosated human serum albumin against multidrug-resistant tumor in animal models. *J Control Release*, 164, (2012), 1-7.

[62] Otagiri, M; Chuang, VT; Maruyama, T; Kragh-Hansen, U. Human Serum Albumin: New insights on its structural dynamics, functional impacts and pharmaceutical applications, Sojo University Publishing Center, Kumamoto, (2013).

[63] Jain, RK. Lessons from multidisciplinary translational trials on anti-angiogenic therapy of cancer. *Nat Rev Cancer*, 8, (2008), 309-316.

[64] Vredenburgh, JJ; Desjardins, A; Herndon, JE 2nd, Dowell, J.M., Reardon, D.A., Quinn, J.A., Rich, J.N., Sathornsumetee, S., Gururangan, S., Wagner, M., Bigner, D.D., Friedman, A.H., Friedman, H.S. Phase II trial of bevacizumab and irinotecan in recurrent malignant glioma. *Clin Cancer Res*, *13*, (2007), 1253-1259.

[65] Clark, AJ; Lamborn, KR; Butowski, NA; Chang, SM; Prados, MD; Clarke, JL; McDermott, MW; Parsa, AT; Berger, MS; Aghi, MK. Neurosurgical management and prognosis of patients with glioblastoma that progresses during bevacizumab treatment. *Neurosurgery*, 70, (2012), 361-370.

[66] Hu, YL; DeLay, M; Jahangiri, A; Molinaro, AM; Rose, SD; Carbonell, WS; Aghi, MK. Hypoxia-induced autophagy promotes tumor cell survival and adaptation to antiangiogenic treatment in glioblastoma. *Cancer Res*, 72, (2012), 1773-1783.

[67] Mazure, NM; Pouysségur, J. Hypoxia-induced autophagy: cell death or cell survival? *Curr Opin Cell Biol*, 22, (2010), 177-180.

[68] Leone, RD; Amaravadi, RK. Autophagy: a targetable linchpin of cancer cell metabolism. *Trends Endocrinol Metab*, 24, (2013), 209-217.

[69] Tracy, K; Dibling, BC; Spike, BT; Knabb, JR; Schumacker, P; Macleod, KF. BNIP3 is an RB/E2F target gene required for hypoxia-induced autophagy. *Mol Cell Biol*, 27, (2007), 6229-6242.

[70] Hu, YL; Jahangiri, A; Delay, M; Aghi, MK. Tumor cell autophagy as an adaptive response mediating resistance to treatments such as antiangiogenic therapy. *Cancer Res*, 72, (2012), 4294-4299.

[71] Ishima, Y; Inoue, A; Fang, J; Kinoshita, R; Ikeda, M; Watanabe, H; Maeda, H; Otagiri, M; Maruyama, T. Poly-S-nitrosated human albumin enhances the antitumor and antimetastasis effect of bevacizumab, partly by inhibiting autophagy through the generation of *nitric oxide*. *Cancer Sci*., 106, (2015), 194-200.

[72] Ogasawara, Y; Ishii, K; Togawa, T; Tanabe, S. Determination of bound sulfur in serum by gas dialysis/high-performance liquid chromatography. *Anal Biochem*, 215, (1993), 73–81.

[73] Ogasawara, Y; Isoda, S; Tanabe, S. Tissue and subcellular distribution of bound and acid-labile sulfur, the enzymic capacity for sulfide production in the rat. *Biol Pharm Bull*, 17, (1994), 1535–1542.

[74] Toohey, JI. Sulphane sulphur in biological systems: a possible regulatory role. *Biochem J*, 264, (1989), 625–632.

[75] Bordo, D; Bork, P. The rhodanese/C25 phosphatase superfamily. Sequence-structurefunction relations. *EMBO Rep*, 3 (2002), 741–746.

[76] Vennesland, BV; Castric, PA; Conn, EE; Solomonson, LP; Volini, M; Westley, V. *Fed. Proc. Fed. Am. Soc. Exp. Biol.*, 41, (1982), 2639-2648.

[77] Wintner, EA; Deckwerth, TL; Langston, W; Bengtsson, A; Leviten, D; Hill, P; Insko, MA; Dumpit, R; VandenEkart, E; Toombs, CF; Szabo, C. A monobromobimane-based assay to measure the pharmacokinetic profile of reactive sulphide species in blood. *Br J Pharmacol.*, 160, (2010), 941–957.

In: Human Serum Albumin (HSA)
Editor: Travis Stokes

ISBN: 978-1-63482-963-2

Chapter 2

Human Serum Albumin (HSA) Drug Delivery System

Dorota Bartusik[1,•], David Aebisher[2] and Boguslaw Tomanek[3]

[1]Southern Polytechnic State University, Department of Biology and Chemistry, Marietta, Georgia, US
[2]Shorter University, Natural Sciences Department, Rome, Georgia, US
[3]University of Alberta, Department of Oncology, Edmonton, Alberta, Canada

Abstract

Human Serum Albumin (HSA) has been widely used for drug delivery systems. HSA binds to variety of drugs resulting in HSA-drug complexes. HSA influences the pharmacokinetic and pharmacodynamic properties of drugs and the complexes have demonstrated potential for improving drug tracking and efficiency.

This review discusses the properties of drug binding sites within the structure of HSA. We analyzed recently reported HSA-drug complexes including novel HSA-antibody conjugates. According to the literature reviewed, we present an overview of the HSA-drug complexes that may potentially lead to significant clinical applications by improving drug efficiency and control of distribution and release rate of drug. Moreover, we discuss the fabrication of HSA drug delivery systems and the factors that optimize particle sizes and distribution and the effect of pH on the HSA-drug complexes. Also, we discuss delivery of HSA-anticancer drug complexes to receptors expressed on the cancer cells surfaces. Research concerning *in vitro* and *in vivo* techniques used for the HSA-drug complex monitoring is also presented in this chapter.

Introduction

Drug delivery systems are systems developed to improve the process of drug administration. The development effective drug delivery systems are one of the most

• E-mail: dorota.bartusik@gmail.com.

challenging tasks in pharmacy. Ongoing research continues to develop systems which can carry drugs and deliver them to targets. With this approach, many drugs in use currently can be incorporated into more efficient drug delivery systems. Historically, the first drug delivery systems developed were liposome nanoparticles in 1974 and this early work lead to several discoveries using nanoparticles as drug carriers (Gregoriadis 1974).

In this review, we give an overview of the drug delivery systems that make use of Human Serum Albumin (HSA) as a drug carrier. Albumin-based drug delivery systems represent a promising approach to more specific and tolerable treatments of tissues. HSA an attractive strategy for active monitoring of drugs and its therapeutics efficacy in preclinical studies *in vitro* and clinical studies *in vivo* (Ni 2014; Kratz 2008; Kratz 2014; Müller- Radner 1998). Protein-based drug delivery systems such as HSA are defined as biodegradable and biocompatible carriers of drugs. The interactions between drug delivery systems and albumin in human body are crucial for the efficacy of treatments. HSA is the smallest and most abundant protein in human blood plasma which binds to large number and variety of compounds. HSA affects drug efficiency by affecting drug transport across cell membranes. It has been found that HSA is present in the extravascular and intravascular space.

The studies confirmed that only 40% of the total amount of HSA is present in plasma and 60% is located in extracellular space, muscle, skin and the lymphatic system (Peters 1996; Evans 2002; Fasano 2005; Merlot 2014).

Recent advances in drug synthesis provide the ability to use HSA as a drug carrier to improve the delivery of the drug or prodrug to the site of action (Zhao 2014; Peralta 2014; Sivertsen 2014; Popov 2005, Zharov 2005; Schwartz 1995). Current biomedical research is using HSA conjugates which include HSA carrier forms such as drug nanoparticles, antibody, peptides, free radicals and synthetic polymers. Therefore, targeted drug delivery can be used to treat many diseases including cancerous tumors, rheumatoid arthritis, infectious diseases, cardiovascular disease and diabetes (Zaloga 2014; Yu 2014; Bucova 2008; Carmeliet 2000; Oettinger 2004; Pandey 2004; O'Dell 1997).

In the HSA-binding drug delivery system, drugs are bound to a linker through a cleavable bond. Drugs often contain the useful functional groups used to bond to HSA including but not limited to hydroxyl, carbonyl, aldehyde, haloform, carbonate ester, carboxyl, hydroperoxy, carboxamide and amine (Chung 2006; Paulos 2004; Burish 1992; Chen 1974). If the drug does not contain a suitable functional group, then it may be modified synthetically to introduce a functional group which is able to connect to a suitable linker. Albumin-based drug delivery system comprising an albumin-binding group, a drug, and a linker that can be cleaved hydrolytically or enzymatically (Loganathan 2014; Papazisis 2004; Rihowa 2000; Richieri 1993).

Many types of drug delivery systems are available which enable selection of an appropriate system for specific treatments (Sajid 2015; Shahabadi 2015; Nilverband 2014; Nohynek 2007: Smid 2007; Kaul 2003; Petipas 2001; Nicholson 2000; Neuzil 1994; Peters 1985; Reed 1977). The selection of drug for the drug delivery systems is based on the drug type, the disease state and the location of the target in the body. Small molecule drugs act as active units in drug delivery systems because of small molecular size, permeability, diffusion and solubility (Rahnama 2015; Zheng 2014; Pap 2000; Peters 1996; Neuzil 1994; Niwa 1990; Peters 1985). A key element for the use of suitable drug delivery systems is that the diameter of the smallest capillaries in the human body is on the order of 4 μm. In order to access all locations in the body drug delivery system should preferably have a small diameter (Kamath

2014). Among the variety of drug delivery systems, the choice is dependent upon the bioacceptability of the carrier, drug potential irritant reactions, the particle size and physicochemical properties of the carrier (Junutula 2006; Kaul 2003; Fernandez –San 2003; Cui 1996; Davies 1996; Peters 1985). The number of publications on controlled drug binding over the years 1966-2014 is presented in Figure 1.

The data below (Figure 2) reflects the use of albumin as a drug delivery carrier. The number of articles describing albumin drug delivery systems is increasing rapidly since the first article was published in 1965. Among the increasing number of albumin drug delivery articles, many have been published with title keywords "albumin drug carrier" (Figure 3). The data in figures 1, 2 and 3 were collected from the PubMed database.

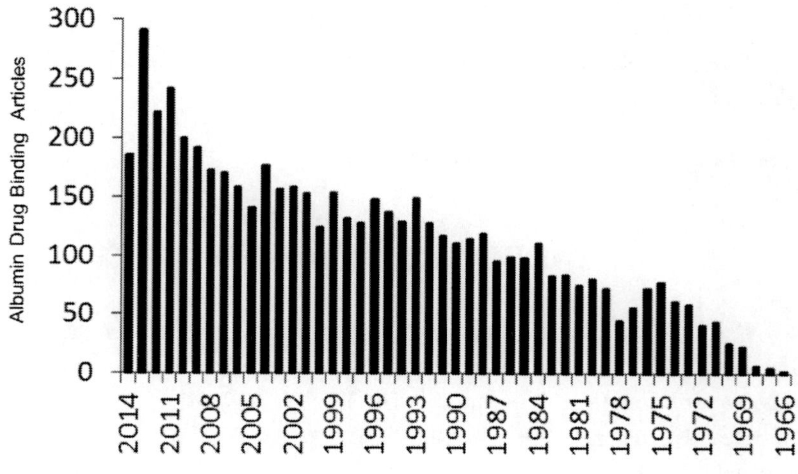

Data collected using PubMed database 2015.

Figure 1. Albumin Drug Binding Articles during 1966-2014.

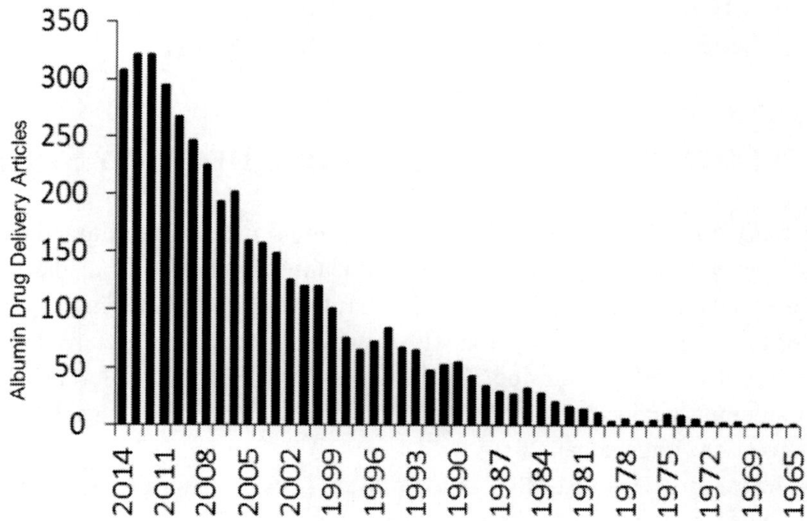

Data collected using PubMed database 2015.

Figure 2. Albumin Drug Delivery Articles during 1965-2014.

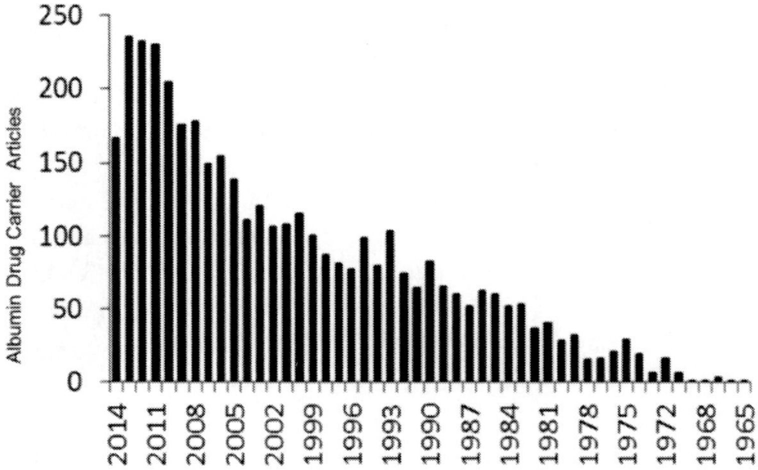

Data collected using PubMed database 2015.

Figure 3. Albumin Drug Carrier Articles during 1965-2014.

THE STRUCTURE OF HSA

In the past 40 years, the binding sites for therapeutic drugs have been studied extensively (Fasano 2005; Bertucci 2002). HSA is composed of three similar globular domains, each of which contains two subdomains, denominated IA, IB, IIA, IIB, IIIA, and IIIB (Sudlow 1976; Hawkins 1982; He 1992; Peters 1996; Sugio 1999).

The two principal regions responsible for ligand-binding to HSA are known as Sudlow's Site I and II, located in subdomain IIA and IIIA (Xu 2015; Sudlow 1976; Peters 1996). HSA has a secondary structure that is 55% α-helix and 45% is believed to be ß-turn. The three-dimensional shape of HSA is described as an ellipsoid consisting of three flexible spheres (domains I, II, III) and has been elucidated by X-ray structure analysis (Yeggoni 2014; Hawkins 1982; Carter 1989; Carter 1994; Peters 1996; Koh 1979).

RECENT DEVELOPMENTS IN HSA DRUG DELIVERY SYSTEMS

Despite many efforts, the therapeutic effect of many drugs is often limited as a result of poor pharmacokinetics, mainly as a consequence of fast elimination from the body (Lewis 2008; Liberatore 1992; Levick 1981; Koh 1979). Drug delivery systems can be developed by coupling of drugs to albumin, conjugation with bioactive proteins and encapsulation of drugs into albumin (Rahnama 2015; Totoribe 2014; Anhorn 2008; Bird 2004; Bertucci 2002). Drug release from albumin nanoparticles or HSA drug delivery systems can be achieved naturally by protease digestion (Loganathan 2014).

To date, research shows that HSA is a multifunctional protein that is able to bind and transport numerous endogenous and exogenous compounds (Merlot 2014; Kitamura 2013; Comisso 2012; Mendez 2005; Martin 2004; Monsour 2003; Mitzner 2001; Christensen 2002; Meden 1997; Moreland 1997; Matsumura 1986; Chen 1974). In the last decade, a variety of applications for HSA have been developed. HSA has a strong negative charge (Coppieters

2006; Nicholson 2000; Sugio 1999) and can bind to acidic and neutral drugs. The cellular membranes which can bind to HSA drug delivery systems are glycoprotein 60, 30, 18 (Schnitzer 1988; Ghinea 1988; Zhai 2000; Amsellem 2010), the neonatal Fc receptor (Andersen 2012; Roopenianand 2007; Anderson 2006; Chaudhury 2003) and heterogeneous nuclear Ribonucleoproteins. HSA is a blocking agent and protein binding agent, and is added to cell culture media as a supplement to increase growth and deliver important nutrients to cells (Totoribe 2014; Coskun 2013; Stehle 1999; Stehle 1997).

Accordingly, our criteria for the selection of papers for this review were focused on albumin-based drug delivery systems which include a therapeutically or diagnostically active drug or chemical agent (Jose 2015; Anjomshoa 2014; Kirak 2014; Cui 2013; Barok 2011; Kreuter 2007; Asghar 2003; Ríhová 2000; Hartung 1999; Stehle 1999; Barton 1990; Babson 1954). Our review demonstrates that the major goals are to design HSA carriers which include nanoparticles (Mukherjee 2014; Peralta 2014; Yu 2014; Kawakami 2000), nanospheres (Kratz 1997, Kundu 2015), nanocapsules (Israel 2015; Kim 2003), superparamagnetic particles of iron oxide (Wacker 2014), gold particles (Miller 2003), fluorescent particles (Kundu 2015), lauric acid/albumin hybrid iron oxide nanoparticle (Zaloga 2014), micelles (Varshosaz 2013; Zhao 2014), fenretinide complex with human serum (Durante 2014), *in situ* gel containing curcumin-loaded albumin nanoparticles (Lou 2014; Azimi 2014), nanocrystal formulation (Lu 2014), copolymer (Chen 2014), fluorescent particles (Wang 2014), gelatin particles (Azimi 2014) and carrier for optical imaging (Li 2014). All these albumin nanocarriers are biodegradable, easy to prepare, and have well-defined sizes and reactive functional groups on their surface that can be used for ligand binding and other surface modifications (Shahzad 2014; Jiang 2014; Merlot 2014; Digilio 2014). Research on development of HSA based drug delivery systems is rapidly growing due to their great therapeutic effects (Jenita 2014; Brekken 2001; Bosse 2005; Yoshikawa 2000; Birn 2000; Coester 2000; Galántai 2000; Ito 2000; Kaye 1998) achieved by the fact that albumin effectively incorporates drugs electrostatically. Charged drug molecules are readily adsorbed because of the high content of charged amino acids in albumin. The presence of functional groups on the nanoparticle surface makes it possible to conjugate targeting ligands. Albumin microcapsules were also developed for delivering antiinflammatory drugs (Oettinger 2012).

A number of relevant studies present the development of microspheres produced by chemically linking albumin using glutaraldehyde and stabilization at high temperatures (Gayakwad 2009). The size of the microspheres was usually in a range of 1-100 μm. The drug was dispersed through the albumin matrix and released continuously as the microcapsules were degraded by intracellular enzymes (Gayakwad 2009). After encapsulating an antisense NF-kappaB oligonucleotide in HSA drug delivery systems, the albumin microsphere could effectively reduce paw swelling in a rat rheumatic arthritis model (Moreland 1997). The most commonly used drug aspirin, was incorporated into particle size ranging from 47-191 nm with an aspirin/albumin ratio of 0.06-1.0. Aspirin was released from the nanoparticle at a sustained rate for a prolonged duration with 50% total cumulative release at the end of 20 hours (Das 2005). The use of albumin drug systems significantly prolonged the serum half-life and promoted targeting of inflamed joints (Liu 2012; Coppieters 2006; Schiff 2000; Keyszer 1995;). Folic acid-functionalized HSA have also been developed (Thomas 2011; Turk 2002) to improve drug delivery in rheumatic diseases (Paulos 2004; van der Heijden 2009; Breedveld 1989; Ballantyne 1971). The next generation of nanocapsules with 440 nm folic acid-functionalized HSA nanocapsules were developed by Rollett and coorkers (Rollett 2012). A 99mTc-labeled HSA has been used in nuclear medicine for diagnosing various conditions including cancer and infectious diseases (Adams 2001; Rink 2001; Wang 2007). Many drug delivery systems and formulation are available such as 99mTc-Nanocoll® and 99mTcAlbures®. In these formulations, the human serum albumin aggregated particles has a wide range of size from a few nm to up to 1000 nm depending on the application (Breedveld 1989). Moreover, a secreted albumin-binding protein known as secreted protein, acidic andrichin cysteine has been identified (Xu 2012; Trist 2012; Shenoy 2005; Schuimaker 1996; Schnitzer 1992).

As discussed, currently approved HSA drug delivery systems provide stable formulations, provide improved pharmacokinetics, and a degree of 'passive' or 'physiological' targeting to tumor tissue (Shahzad 2014; Sivertsen 2014; Warnecke 2007; Turk 2002; Simoes 2000; Hartung 1999). Improving efficacy ratios with currently existing drugs can be attempted using more efficient methods for their delivery. HSA binds to toxins to avoid adverse effects; binds to excessive proteins to act as a buffer and binds hormones and growth peptides to keep them stable. HSA can greatly reduce contamination binding that can interfere in common molecular biology assays. It is also used as a blocking agent in many other clinical applications including diagnostic procedures, medical and surgical devices. In addition, HSA is also used in drug delivery to transport pharmaceutical drugs like antibiotics and anti-inflammatories. Since HSA is a physiological carrier of fatty acids, careful consideration must be given to the fact that albumin can deliver impurities to cells and can extract fatty acids and other nutrients from cells, which could alter their metabolic state (Fasano 2005). In addition, there is continuing disagreement about the rates of fatty acids dissociation from albumin and whether this affects biophysical measurements of fatty acids transport (Fanali 2012; Sudlow 1975). Bilirubin combines reversibly with human albumin in neutral or alkaline solutions At least two molecules of the ligand bilirubin can be bound to one HSA protein. The kinetics of this process in both directions, and equilibrium states has been studied by several groups (Gray 1978; Reed 1977; Chen 1974). Bilirubin is a waste product from the breakdown of red blood cells. The liver processes bilirubin so it can be excreted in stool. Bilirubin flows through the liver's bile ducts, dissolved in bile. The process of binding anticancer drugs with the HSA was studied by a combination of spectroscopic techniques and molecular docking method. These interactions resulted in the formation of drug-HSA complex, making the local microenvironment in subdomain IIA of the protein more hydrophobic than its native state. The results indicated that the binding interaction were van der Waal's forces and hydrogen bonding interactions (Xiaoqing 2015). The accumulation of albumin in solid tumors forms the rationale for developing albumin-based drug delivery systems for tumor targeting (Azimi 2014; Hasan 2013; Kratz 2010; Desai 2009; Kratz, 2008; Asghar 2003; Babson 1954). Other groups have developed and reported the interaction of novel polyvinylthiol coated silver nanoparticles with HSA studied by fluorescence, UV–visible, and circular dichroism spectroscopies (Ali 2015). Recent studies on HAS systems include doxorubicin prodrugs that are cleaved by matrix metalloproteases 2 and 9 cathepsin B (Graeser 2008; Anhorn 2008; Schmid 2007), urokinase (Chung 2006) or prostate-specific antigen (Kratz 2006; Greaser 2008) methotrexate prodrugs that are cleaved by cathepsin B and plasmin (Warnecke 2007; Hartung 1999) and camptothecin prodrugs that are cleaved by cathepsin B or unidentified proteases (Schmid 2007), nab-paclitaxel (Sun 2014; Gupta 2014), dexketoprofen (Lu 2014), resveratrol (Kamath 2014), oral antidiabetic nanomedicines

(Gundogdu 2014), fluorine-18 radio-labeled serum albumins (Falguni 2015) and dexatropofen (Lu 2014). Albumin nanoparticles are one of the most important drug carriers for the delivery of therapeutic drugs, especially for the treatment of malignancies (Kouchakzadeh 2015; Anjomoshoa 2014). Paclitaxel - loaded albumin nanoparticle, *nab*-paclitaxel or Abraxane®, is already on the market and was initially approved for clinical use in the United States in 2005 (Kudlowitz 2014) for the treatment of metastatic breast cancer (Kudlowitz 2014). More recently, Abraxane® has also been described for the treatment of metastatic adenocarcinoma of the pancreas, in combination with gemcitabine, and metastatic non-small cell lung carcinoma, in combination with carboplatin (Kudlowitz 2014). However, studies have also demonstrated mechanisms by which albumin acts to effectively improve therapeutic use of drugs and reduce rapid clearance (Souza 2014; Merlot 2014Tishler 2010; Steinhauser 2006; Dennis 2002; Yuan 1995).

Thiosemicarbazone, namely 2-pyridylketone4,4-dimethyl-3-thiosemicarbazone (Merlot 2013), has been shown to be internalized by cancer cells receptors (Merlot 2014). The uptake, toxicity and apoptotic activity of thiosemicarbazone is greatly enhanced in the presence of HSA (Merlot 2014). Specific examples of albumin-based drug delivery systems include also docetaxel or rapamycin (Kratz 2008).

Currently, there are numerous techniques that are used for cancer treatment (Burish 1992; Schwartz 1995). Chemotherapy is the treatment of cancer with anticancer drugs. Another technique of cancer treatment is radiation therapy, which uses radiation energy to destroy cancer cells and reduce the size of tumors (Steeland 2015; Sun 2014; Tormey 1996; Barton 1990; Tormey 1986). A summary of the recent approaches of laser applications includes the laser use to shrink or destroy tumors. In this strategy, laser therapy is most commonly used to treat tumors on the surface of the body or the lining of internal organs. Photodynamic therapy is a type of research that uses a drug called a photosensitizer or photosensitizing agent for the cancer treatment (Schuitmaker 1996). Recent research work has been focused on studying gene therapy for cancer treatment (Durrant 2003). Some of the applications of gene therapy is an experimental treatment that involves introducing genetic material into the cancer cells to destroy the cell. In addition to these techniques, hyperthermia (also called thermal therapy or thermotherapy) is a type of cancer treatment technique in which the cancer cells are exposed to high temperatures (up to 113◦F) (Wust 2002). Research has shown that high temperatures can damage and kill cancer cells with minimal injury to normal tissues. By damaging proteins and functional structures within cells, hyperthermia destroys cancer cells (Wust 2002).

Drug delivery systems can not only protect active pharmaceutical ingredients against degradation, but also offer a prolonged release of drugs in a controlled reaction. Solid tumors commonly possess an immature and highly permeable vasculature (Carmeliet 2000; Maeda 2000; Greish, 2007; vander Veldt, 2008), however, despite this, there is generally insufficient lymphatic drainage (Carmeliet 2000; Maeda 2000; Greish, 2007) and results in an accumulation of macromolecules (>40 kDa) within the tumor interstitium due to enhanced permeation and retention effect (Maeda 2000; Greish 2007). Additionally, a number of studies have proposed that tumors are a site of albumin catabolism (Hradec 1958; Andersson 1991; Schilling 1992; Stehle 1997). The example of anticancer drug conjugated to albumin was a methotrexate. This conjugate, methotrexate-HSA conjugate (MTX-HSA), was synthesized by directly coupling the drug to lysine residues of HSA (Hartung 1999). Other examples of HSA drug delivery systems include β-Carotene and astaxanthin, two carotenoids with powerful antioxidant properties (Li 2015). In this study, the interaction of these two carotenoids with HSA under physiological conditions was investigated using several spectroscopic techniques. The experimental results indicate the quenching mechanism of HSA by the two carotenoids, is a static process (Li 2015).

The administration of the individual albumin- based drug delivery systems may be in the range of a few minutes to hours. The cleavable linker may be cleavable by an enzyme. For example, the cleavable linker may contain at least one peptide bond which preferably lies within a cleavable peptide sequence of a protease. A peptide bond can therefore be implemented by the insertion of a respective peptide sequence into the cleavable linker. Suitable enzymes are, for example, proteases (cysteine proteases, serine proteases) and peptidases, which include matrix metalloproteases, and plasmin activators, which are formed or activated in intensified manner in diseases such as rheumatoid arthritis or cancer, leading to excessive tissue degradation, inflammations and metastasis (Loganathan 2014; Papazisis 2004; Rihowa 2000; Richieri 1993). Examples of proteases suitable to drug delivery systems are in cathepsin, plasmin, urokinase, and prostate-specific antigens. For most applications the development of albumin as a drug carrier is increasingly important to consider in terms of the targeted delivery of cancer therapy (Li 2014; Cui 2013; Coester 2000). Albumin has several advantages over synthetic polymers, as it is biocompatible, biodegradable, has low cytotoxicity and has an excellent binding capacity with various drugs (Yu 2014; Van der Heijden 2009; Yoshikawa 2000). Since existing tumor chemotherapeutic agents neither target tumour cells, nor are they specific to tumor cells, a slow release of drugs from carriers would be beneficial in targeting carcinogenic cells intracellularly (Jagoda 2015; Shahzad 2014; Dinauer 2005; El Sayed 2005; Fayette 2005; El Aneed 2004; Endo 1987).

Many studies have reported the use of antibodies to develop drug delivery systems. Antibodies can improve the therapeutic efficacy of the drug substances as well as improve distribution. Various drug delivery systems have been investigated to treat one widely expressed receptor which is human epidermal growth factor receptor-2 (Her-2) which is a member of the Her-2 family EGFR (ErbB-1), Her-2 (ErbB-2), Her-3 (ErbB-3) and Her-4 (ErbB-4) (Harris 1992; Davies 1996). One of the most popular antibodies is Trastuzumab (more commonly known under the trade name Herceptin), a humanized IgG1 monoclonal antibody that acts on human Her2/neu receptors. In 1998 Herceptin was approved by the United States Food and Drug Administration (FDA) for the treatment of metastatic breast cancer. In 2010 Herceptin was also approved by the FDA for the treatment of metastatic cancer of the stomach. However, the problems commonly associated with treating drug and used in breast cancer treatment is tamoxifen. Poly(ethylene oxide)–modified poly(ε-caprolactone) polymeric nanoparticles have been tested for their target specificity as a carrier for tamoxifen (Shenoy 2005).

Research studies done on monoclonal antibodies attached to HSA shows that thiolation with 100-fold molar excess of 2-iminothiolane and the ratio of 10:1 for the thiolated 1F2 (μg) to PEGylated nanoparticles (mg), were optimum for the attachment to Her2 receptor on the surface of BT474 cells (Kouchakzadeh 2013; Nilvebrant 2014).

Studies done by Wan and coworkers prove that in 4T1 breast cancer cells, HSA nanoparticles loaded with lapatinib can and enhance lapatinib efficacy (Wan 2015). Lapatinib is a selective small-molecular dual-tyrosine kinase inhibitor of Her2 and EGFR, and is effective in reducing cancer progression and metastasis. Pharmacokinetic studies demonstrated that HSA nanoparticles with Laptanibab effectively increased the accumulation of drug in breast tumor tissue (Wan 2015). Recent research done by Xu and coworkers (Xu 2015), has been focused on the study of the binding interaction of 5-fluorouracil to HSA in subdomain IIA of the HSA molecule. These studies used a combination of spectroscopic techniques and molecular docking methods to confirm drug efficiency (Xu 2015).

Another study was designed to examine the interaction of Tenofovir with HSA. The results confirmed that Tenofovir interacted with HSA and that hydrophobic interaction plays a major role in the Tenofovir-HSA complex. Through the site marker competitive experiment, Tenofovir was confirmed to be located in site I of HSA (Shahabadi 2015). Work done using capecitabine was focused on investigations of treatments of advanced human cancers, especially gastrointestinal tract, ovary, colorectal and breast cancers. Drugs and drug metabolites are transported in the blood by transporter proteins like HSA (Ishtikhar 2014; Garg 2014; Goldmacher 2011). Also pertuzumab was recently investigated as a potential drug for HER2 target analysis (Shahzad 2014). The hydrophobic and hydrophilic interactions when piperine binds to HSA were studied by Yeggoni and coworkers (Yeggoni 2014). HSA nanoparticles represent an attractive strategy for active targeting of therapeutics into tumor cells due to the presence of superficial functional groups. Novel approaches of the use of HSA describe monoclonal antibodies-HSA drug delivery systems. The preparation of these monoclonal antibodies-HSA conjugates involves several reaction steps including the reduction of one reaction partner to give a free thiol group and the introduction of maleimide groups to the complementary protein prior to the actual coupling reaction.

The use of HSA nanoparticles in drug delivery systems represent an attractive strategy for active targeting of therapeutics into tumor cells due to the presence of superficial functional groups. Novel approaches of the use of HSA describes monoclonal antibodies - HSA drug delivery systems. A Series of small linear cyclictetrapeptides containing a β 2,2-amino acid with either two 2-naphthyl-methylene or two para-CF3-benzyl side chains, along with their interaction with the main plasma protein HSA have been studied (Sivertsen 2014). The preparation of monoclonal antibodies -HSA conjugates involves several reaction steps including the reduction of one reaction partner to give free thiol groups and the introduction of maleimide groups to the complementary protein prior to the actual coupling reaction.

The strategies for development of HSA drug delivery systems can be distinguished using three ways. Firstly, antibodies or peptide are conjugated to an albumin-binding domain, secondly, the Fc region is substituted by albumin and two antigen-binding regions (Fab regions) are fused to albumin; and thirtly an albumin-binding peptide is bound to Fab regions, other antibody fragments, or therapeutic peptides that binds to serum albumin with high affinity in the nano- to fentomolar range (Wan 2015; Yegoni 2014;Varshoaz 2013). The feature is shared with the IgG class of antibodies, that both albumin drug delivery systems with monoclonal antibodies bind to a cellularly expressed receptor, named the neonatal Fc receptor (FcRn). They bind to different binding sites, may bind FcRn simultaneously, and do so in a non-cooperative manner. Binding to both ligands is strictly pH dependent, with binding at acidic pH (pH = 6.0) and no binding or release at physiological pH. Both binding interfaces contain conserved histidine residues, located to the IgG Fc, domain III (DIII) of albumin as well as the receptor, which becomes positively charged at acidic pH. Ligand binding at acidic pH results in transport back to the cell surface, where exposure to the neutral pH of the blood triggers release of IgG and albumin back into the circulation (Wan 2015; Yegoni 2014; Wee 2005; Wust 2002; Yoo 1999). The receptor-mediated recycling mainly takes place in hematopoietic cells and endothelial cells lining the vascular space. The pore size of tumor microvessels varies from 100 to 1200 nm while HSA has an effective diameter of 7.2 nm allowing extravasation into tumor tissue but not into normal tissue (Wang 2005; We 2005Yoshikawa 2000; Weiner 1994).

SUMMARY

Our review presents trends and technologies in the field of HSA drug delivery systems. Various studies related to formulation and characterization of drug delivery systems were discussed herein. The research in drug delivery and targeting mostly represents formation of HSA nanoparticles or nanospheres for the treatment of cancer. With an increasing interest in drug delivery systems there is a greater need to identify biomaterials that are biocompatible and safe for human applications. To this end, protein polymers from animal and plant sources are promising materials for designing nanocarriers (Wacker 2014; Voura 2004).

Human serum albumin nanoparticles were used as drug delivery systems, and applied to cancer treatment due to stability and activity of these systems. The high solubility of albumin makes it an attractive carrier, capable of accommodating a wide variety of drugs. A large number of drugs and prodrugs are known to bind to serum albumin, and hence albumin appears to be a promising drug carrier. The use of drug delivery system can improve the efficacy of drugs and prodrugs.

REFERENCES

Adams, B. K.; Al Attia, H. M.; Khadim, R. A.; Al Haider, Z. Y. 99Tc(m) nanocolloid scintigraphy: a reliable way to detect active joint disease in patients with peripheral joint pain. *Nuclear Medicine Communications,* 2001; 22: 315-318.

Amsellem, S.; Gburek, J.; Hamard, G.; Nielsen, R.; Willnow, T. E.; Devuyst, O. Cubilinis essential for albumin reabsorption in the renal proximaltubule. *Journal of the American Society of Nephrology*, 2010; 21:1859–1867.

Andersen, J. T.; Dalhus, B.; Cameron, J.; Daba, M. B.; Plumridge A.; Evans L. Structure-based mutagenesis reveals the albumin-binding site of the neonatal Fc receptor. *Nature Communications*, 2012; 3(611) 1-5.

Anderson, C. L.; Chaudhury, C.; Kim, J.; Bronson, C. L.; Wani, M. A.; Mohanty, S. Perspective - FcRn transports albumin: relevance to immunology and medicine. *Trends in Immunology*, 2006; 27: 343–348.

Andersson, C.; Iresjo, B. M.; Lundholm K. Identification of tissue sites for increased albumin degradation in sarcoma-bearing mice. *Journal of Surgical Research,* 2012; 50:156–162.

Anhorn, M. G.; Wagner, S.; Kreuter, J.; Langer K.; von Briesen, H. Specific Targeting of HER2 Overexpressing Breast Cancer Cells with Doxorubicin-Loaded Trastuzumab-Modified Human Serum Albumin Nanoparticles. *Bioconjugate Chemistry*, 2008; 19: 2321–2331.

Anjomshoa, M.; Fatemi, S. J.; Torkzadeh-Mahani, M.; Hadadzadeh, H. DNA - and BSA - binding studies and anticancer activity against human breast cancer cells (MCF-7) of the zinc(II) complex coordinated by 5,6-diphenyl-3-(2-pyridyl)-1,2,4-triazine. *Spectrochim Acta A Molecular and Biomolecular Spectroscopy,* 2014;127:511-520.

Asghar, R. B.; Ann, M. D.; Spanel, P.; Smith, D.; Simon, J. D. Measuring transport of water across the peritoneal membrane. *Kidney International,* 2003; 64(5):1911-1915.

Azimi, B.; Nourpanah, P.; Rabiee, M.; Arbab, S. Producing gelatin nanoparticles as delivery system for bovine serum albumin. *Iranian Biomedical Journal,* 2014;18(1):34-40.

Babson, A. L.; Winnick, T. Protein transfer in tumor-bearing rats. *Cancer Research,* 1954; 14, 606-611.

Ballantyne, F. C.; Fleck, A.; Dick, D. C. Albumin metabolism in rheumatoid arthritis. *Annals of the Rheumatic Diseases*, 1971; 30: 265-270.

Barok, M.; Tanner M.; Koninki K.; Isola J. Trastuzumab-DM1 causes tumour growth inhibition by mitotic catastrophe in trastuzumab-resistant breast cancer cells in vivo. *Breast Cancer Research*, 2011; 13: R46.

Barton, C.; Waxman, J. Effects of chemotherapy on fertility. *Blood Review,* 1990; 4:187–195.

Bertucci, C.; Domenici, E. Reversible and covalent binding of drugs to human serum albumin: methodological approaches and physiological relevance. *Current Medicinal Chemistry*, 2002; 9:1463-1481.

Bird, S. M. Recipients of blood or blood products "at vCJD risk". *British Medical Journal,* 2004; 328:118-119.

Birn, H.; Fyfe, J. C.; Jacobsen, C.; Mounier, F.; Verroust, P. J.; Orskov, H.; Cubilinis an albumin binding protein important for renal tubular albuminreabsorption. *Journal of Clinical Investigation*, 2000; 105, 1353–1361.

Breedveld, F. C.; van Kroonenburgh, M. J.; Camps, J. A.; Feitsma, H. I.; Markusse, H. M.; Pauwels, E. K. Imaging of inflammatory arthritis with technetium-99m-labeled IgG. *Journal of Nuclear Medicine*, 1989; 30: 2017-2021.

Brekken, R. A.; Sage, E. H. SPARC,amatricellularprotein: at the crossroads of cell-matrix communication. *Matrix Biology*, 2001;19, 816–827.

Bosse, D.; Praus, M.; Kiessling, P.; Nyman, L.; Andresen, C.; Waters, J.; Schindel, F. Phase I comparability of recombinant human albumin and human serum albumin. *Journal of Clinical Pharmacology*, 2005; 45:57-67.

Bucova, M.; Bernadic, M.; Buckingham, T. C-reactive protein, cytokines and inflammation in cardiovascular diseases. *Bratislavské Lekárske Listy,* 2008; 109(8):333-340.

Burish, T. G.; Tope, D. M. Psychological techniques for controlling the adverse side effects of cancer chemotherapy: Findings from a decade of research. *Journal Pain Symptom Manage*, 1992; 7:287–301.

Carmeliet, P; Jain, R. K. Angiogenesisincancerandotherdiseases. *Nature*, 2000; 407: 249–257.

Carter, D. C.; He, X. M.; Munson, S. H.; Twigg, P. D.; Gernert, K. M.; Broom, M. D.; Miller, T. Y. Three-dimensional structure of human serum albumin. *Science*, 1989; 244:1195–1198.

Carter, D. C.; Ho, J. X. Structure of serum albumin. *Advances in Protein Chemistry*, 1994; 45:153-203.

Chaudhury, C.; Mehnaz, S.; Robinson, J. M.; Hayton,W. L.; Pearl, D. K.; Roopenian, D. C. The majorhis to compatibility complex-related F creceptor for IgG(FcRn) binds albumin and prolongs its life span. *Journal of Experimental Medicine*, 2003; 197, 315–322.

Chen, R. F. Fluorescence stopped-flow study of relaxation processes in the binding of bilirubin to serum albumins. *Archive in Biochemistry and Biophysics,* 1974; 160:106-112.

Chen, X.; Lv, G.; Zhang, J.; Tang, S.; Yan, Y.; Wu, Z.; Su, J.; Wei, J. Preparation and properties of BSA-loaded microspheres based on multi-(amino acid) copolymer for protein delivery. *International Journal of Nanomedicine,* 2014; 9:1957-1965.

Christensen, E. I.; Birn, H. Megalin and cubilin: multifunctional endocyticreceptors. *Nature Reviews Molecular Cell Biology*, 2002; 3: 256–266.

Chung, D. E.; Kratz, F. Development of a novel albumin-binding prodrug that is cleaved by urokinase-type-plasminogen activator (uPA). *Bioorganic and Medicinal Chemistry Letters,* 2006; 16: 5157–5163.

Coester, C.; Kreuter, J.; von Briesen, H.; Langer, K. Preparation of avidin-labelled gelatin nanoparticles as carriers for biotinylated peptide nucleic acid (PNA). *International Journal of Pharmacology*, 2000; 196(2):147-149.

Commisso, C.; Davidson, S. M.; Soydaner-Azeloglu, R. G.; Parker, S. J.; Kamphorst, J. J.; Hackett, S. Macropinocytosis of protein is anaminoacid supply route in Ras-transformed cells. *Nature,* 2012; 497, 633–637.

Coppieters, K.; Dreier, T.; Silence, K.; de Haard, H.; Lauwereys, M.; Casteels, P.; Beirnaert, E.; Jonckheere, H.; Van de Wiele, C.; Staelens, L.; Hostens, J.; Revets, H.; Remaut, E.; Elewaut, D.; Rottiers, P. Formatted anti-tumor necrosis factor alpha VHH proteins derived from camelids show superior potency and targeting to inflamed joints in a murine model of collagen-induced arthritis. *Arthritis and Rheumatology*, 2006; 54: 1856-1866.

Cross, S. E.; Innes, B.; Roberts, M. S.; Tsuzuki, T.; Robertson, T. A.; McCormick, P. Human skin penetration of sunscreen nanoparticles in vitro assessment of a novel micronized zinc oxide formulation. *Skin Pharmacology and Physiology,* 2007; 20:148–154.

Coskun, A. F.; Nagi, R.; Sadeghi, K.; Phillips, S.; Ozcan, A. Albumin testing in urine using a smart-phone. *Lab. on a Chip*, 2013;13(21):4231-4238.

Cui, M.; Naczynski, D. J.; Zevon, M.; Griffith, C. K.; Sheihet, L.; Poventud-Fuentes, I.; Chen, S.; Roth, C. M.; Moghe, P. V. Multifunctional albumin nanoparticles as combination drug carriers for intra-tumoral chemotherapy. *Advanced Healthcare Materials*, 2013;2(9):1236-1245.

Cui, S.; Verroust, P. J.; Moestrup, S. K.; Christensen, E. I. Megalin/gp330 media te suptake of albumin in renal proximal tubule. *American Journal of Physiology,* 1996; 271: F900–F907.

Davies, D. E.; Chamberlin, S. G. Targeting the epidermal growth factor receptor for therapy of carcinomas, *Biochemistry and Biophysics Research Communication,* 1996; (51) 1101-1110.

Das, S.; Banerjee, R.; Bellare, J. Aspirin loaded albumin nanoparticles by coacervation: implications in drug delivery. Trends *in Biomaterials and Artificial Organs,* 2005; 18: 203-212.

Dennis, M. S.; Zhang, M.; Meng, Y. G.; Kadkhodayan, M.; Kirchhofer, D.; Combs, D. Albumin binding as a general strategy for improving the pharmacokinetics of proteins. *Journal of Biological Chemistry*, 2002; 277, 35035–35043.

Desai, N.; Trieu,V.; Damascelli, B.; Soon-Shiong, P. SPARC expression correlate s with tumor response to albumin-bound paclitaxel in head and neck cancer patients. *Translational Oncology*, 2009; 2: 59–64.

Desai, N. P.; Trieu,V.; Hwang, L. Y.; Wu, R.; Soon-Shiong, P.; Gradishar, J. Improved effectiveness s of nanoparticle albumin-bound(nab) paclitaxel versus polysorbate-based docetaxel in multiplexenografts as a function of HER2 and SPARC status. *Anticancer Drugs*, 2009; 19:899–909.

Digilio, G.; Tuccinardi, T.; Casalini, F.; Cassino, C.; Dias, D. M.; Geraldes, C. F.; Catanzaro, V.; Maiocchi, A.; Rossello, A. Study of the binding interaction between fluorinated matrix metalloproteinase inhibitors and Human Serum Albumin. *European Journal of Medicinal Chemistry*, 2014; 79:13-23.

Dinauer, N. B.; Sabine, W.; Carolin, K.; Kreuter, J.; Langer, K.; von Briesen, H. Selective targeting of antibody-conjugated nanoparticles to leukemic cells and primary T-lymphocytes. *Biomaterials,* 2005; 26:5898–5906.

Drevs, J.; Esser, N.; Richly, H.; Skorzec, M.; Scheulen, M. E.; Unger, C.; Kratz, F. In vivo efficacy and pharmacokinetic study of an acid-sensitive albumin conjugate in a murine renal cell carcinoma model. *Clinical Cancer Research,* 2000; 6(120 Suppl.).

Durante, S.; Orienti, I.; Teti, G.; Salvatore, V.; Focaroli, S.; Tesei, A., Pignatta, S.; Falconi M. Anti-tumor activity of fenretinide complexed with human serum albumin in lung cancer xenograft mouse model. *Oncotarget,* 2014; 5(13):4811-4820.

Durrant, L. G.; Scholefield, J. H. Principles of cancer treatment by immunotherapy. *Surgery*, 2003; 21:277–279.

El-Aneed, A. Current strategies in cancer gene therapy. *Eur. J. Pharm*., 2004; 498:1–8.

El-Sayed, I. H.; Huang, X.; El-Sayed, M. A. Surface plasmon resonance scattering and absorption of anti-EGFR antibody conjugated gold nanoparticles in cancer diagnostics: Applications in oral cancer. *Nanotechnology Letters*, 2005; 5:829–834.

Endo, N.; Kato, Y.; Takeda, Y.; Saito, M.; Umemoto, N.; Kishida, K.; Hara, T. *Cancer Research*, 1987; 47: 1076–1080.

Fanali, G.; diMasi, A.; Trezza, V.; Marino, M.; Fasano, M.; Ascenzi, P. Human serum albumin: from bench to bedside. *Moleular Aspects of Medicine,* 2012; 33(3):209-290.

Fasano, M.; Curry, S.; Terreno, E.; Galliano, M.; Fanali, G.; Narciso, P.; Notari, S.; Ascenzi, P. The extraordinary ligand binding properties of human serum albumin. *IUBMB Life*, 2005; 57(12):787-796.

Fernández-San, M. A.; Angel, M. C.; Miller, M.; Daniell, H. A chloroplast transgenic approach to hyper-express and purify human serum albumin, a protein highly susceptible to proteolytic degradation. *Plant Biotechnology Journal,* 2003;1(2):71-79.

Furst, D. E. Practical clinical pharmacology and drug interactions of low-dose methotrexate therapy in rheumatoid arthritis. *British Journal of Rheumatology,* 34:(Suppl. 2):1995; 20.

Fritzsche,T.; Schnolzer, M.; Fiedler, S.; Weigand, M.; Wiessler, M.; Frei, E. Isolation and identification of heterogeneous nuclear ribonucleoproteins (hnRNP) from purified plasma membranes of human tumour cell lines asalbumin-bindingproteins. *Biochemistry in Pharmacology,* 2004; 67, 655–665.

Fayette, J.; Soria, J. C.; Armand, J. P. Use of angiogenesis inhibitors in tumour treatment. *European Journal of Cancer*, 2005; 41:1109–1116.

Galántai, R.; Bárdos-Nagy, I. The interaction of human serum albumin and model membranes. *International Journal of Pharmacology*, 2000;195(1-2): 207-218.

Galántai, R.; Bárdos-Nagy, I.; Módos, K.; Kardos, J.; Závodszky, P.; Fidy, J. Serum albumin-lipid membrane interaction influencing the uptake of porphyrins. *Archives of Biochemistry and Biophysics*, 2000; 373(1):261-270.

Garg, A.; Quartino, A.; Li, J.; Jin, J.; Wada, D. R.; Li, H.; Cortés, J.; McNally, V.; Ross, G.; Visich, J.; Lum, B. Population pharmacokinetic and covariate analysis of pertuzumab, a HER2-targeted monoclonal antibody, and evaluation of a fixed, non-weight-based dose in patients with a variety of solid tumors. *Cancer Chemotherapy Pharmacology*, 2014; 74(4):819-829.

Gayakwad, S. G.; Bejugam, N. K.; Akhavein, N.; Uddin, N. A.; Oettinger, C. E.; D'Souza, M. J. Formulation and in vitro characterization of spray-dried antisense oligonucleotide to NF-kappaB encapsulated albumin microspheres. *Journal of Microencapsulation*, 2009; 26: 692-700.

Gay, S.; Gay, R. E.; W. J. Koopman. Molecular and cellular mechanisms of joint destruction in rheumatoid arthritis: two cellular mechanisms explain joint destruction? *Annual of Rheumatoid. Diseases*, 1993; 52:(Suppl. 1): S39.

Gholipour, N.; Jalilian, A. R.; Khalaj, A.; Johari-Daha, F.; Yavari, K.; Sabzevari, O.; Khanchi, A. R.; Akhlaghi, M. Preparation and radiolabeling of a lyophilized (kit) formulation of DOTA-rituximab with ^{90}Y and ^{111}In for domestic radioimmunotherapy and radioscintigraphy of non-Hodgkin's lymphoma. *Daru,* 2014;22:58.

Goldmacher, V. S.; Kovtun, Y. V. Antibody-drug conjugates: using monoclonal antibodies for delivery of cytotoxic payloads to cancer cells. *Therapeutic Delivery,* 2011; 2: 397–416.

Graeser, R.; Chung, D. E.; Esser, N.; Moor, N.; Schachtele,C.; Unger, C.; Kratz, F. Synthesis and biological evaluation of an albumin-binding prodrug of doxorubicin that is cleaved by prostate-specific antigen (PSA) in a PSA-positive orthotopic prostate carcinoma model (LNCaP), *International Journal of Cancer*, 2008; 122: 1145–1154.

Gray, R. D.; Stroupe, S. D. Kinetics and Mechanism of Biluribin Binding to Human Albumin. *Journal of Biological Chemistry*, 1978; 253:4370-4377.

Gregoriadis, G.; Willis, E. J.; Swain, C. P.; Tavill, A. S. Drug carrier potential of liposomes in cancer chemotherapy. *Lancet,* 1974; 1:1313–1316.

Guha, M. T-DM1 impresses at ASCO. *Nature Biotechnology*, 2012; 30: 728–728.

Gundogdu E., Yurdasiper A. Drug transport mechanism of oral antidiabetic nanomedicines. *International Journal of Endocrinology and Metabolism,* 2014 Jan. 1;12(1):e8984.

Gupta, N.; Hatoum, H.; Dy, G. K. First line treatment of advanced non-small-cell lung cancer - specific focus on albumin bound paclitaxel. *Inernationl Journal of Nanomedicine,* 2014;9:209-221.

Hartung, G.; Stehle, G.; Sinn, H.; Wunder, A.; Schrenk, H. H.; Kränzle, L.; Fiebig, H. H.; Maier-Borst, W.; Heene, D. L.; Queißer, W. Phase I trial of a methotrexatealbumin in a weekly intravenous bolus regimen in cancer patients. *Clinical Cancer Research*, 1999; 5: 753–759.

Hasan Kouchakzadeh, Seyed Abbas Shojaosadati, Fathollah Tahmasebi, Fazel Shokri. Optimization of an anti-HER2 monoclonal antibody targeted delivery system using PEGylated human serum albumin nanoparticles. *International Journal of Pharmaceutics*, 2013; 447:1–2, 62–69.

Hawkins, J. W.; Dugaiczyk, A. The human serum albumin gene: structure of a unique locus. *Gene,* 1982;19:55-58.

He, X. M.; Carter, D. C. Atomic structure and chemistry of human serum albumin. *Nature,* 1992; 358:209–215.

Harris, A. L.; Nicholson, S.; Sainsbury, R.; Wright, C.; Farndon, J. Epidermal growth factor receptor and other oncogenes as prognostic markers. *Journal of National Cancer Institute,* 1992; (11):181-187.

Herman, R. A.; Veng Pedersen, P.; Hoffman, J.; Koehnke, R.; Furst, D. E. Pharmacokinetics of low-dose methotrexate in rheumatoid arthritis patients. *Journal of Pharmaceutical Sciences*, 1989; 78:165.

Hildebrandt, B.; Wust, P.; Ahlers, O. The cellular and molecular basis of hyperthermia. *Crit. Review Oncology and Hematology*, 2002; 43:33–56.

Hořejší, V.; Drábe, P. Cross-reactivity between tubulin and denatured human serum albumin demonstrated by monoclonal antibody TU-01. *Immunology Letters*, 1984; 8 (5): 285-288.

Ishtikhar, M.; Rabbani, G.; Khan, R. H. Interaction of 5-fluoro-5'-deoxyuridine with human serum albumin under physiological and non-physiological condition: a biophysical investigation. *Colloids and Surfaces B: Biointerfaces,* 2014;123:469-477.

Israel, L. L.; Kovalenko, E. I.; Boyko, A. A.; Sapozhnikov, A. M.; Rosenberger, I.; Kreuter, J.; Passoni, L.; Lellouche, J. P. Towards hybrid biocompatible magnetic Human serum albumin-based nanoparticles: use of ultra-small (CeLn)3/4+ cation-doped maghemite nanoparticles as functional shell. *Nanotechnology*, 2015; 26(4):045601.

Ito, D.; Ogasawara, K.; Iwabuchi, K.; Inuyama, Y.; Onoé, K. Induction of CTL responses by simultaneous administration of liposomal peptide vaccine with anti-CD40 and anti-CTLA-4 mAb. *Journal of Immunology,* 2000; 164(3):1230-1235.

Jain, K. K. Nanotechnology-based drug delivery for cancer. *Technology Cancer Research and Treatments,* 2005; 4:407–416.

Jenita, J. L.; Chocalingam, V.; Wilson, B. Albumin nanoparticles coated with polysorbate 80 as a novel drug carrier for the delivery of antiretroviral drug-Efavirenz. *International Journal of Pharmaceutical Investigation,* 2014; 4(3):142-148.

Jiang, Y.; Lu, H.; Khine, Y. Y.; Dag, A.; Stenzel, M. H. Polyion complex micelle based on albumin-polymer conjugates: multifunctional oligonucleotide transfection vectors for anticancer chemotherapeutics. *Biomacromolecules,* 2014; 15(11):4195-205.

Jose, P.; Sundar, K.; Anjali, C. H.; Ravindran, A. Metformin-Loaded BSA Nanoparticles in Cancer Therapy: A New Perspective for an Old Antidiabetic Drug. *Cell Biochemistry and Biophysics*, 2015; 71(2):627-636.

Junutula, J. R.; Flagella, K. M.; Graham, R. A.; Parsons, K. L.; Ha, E. Engineered thio-trastuzumab-DM1 conjugate with an improved therapeutic index to target human epidermal growth factor receptor 2-positive breast cancer. *Clinical Cancer Research*, 2006; 16: 4769–4778.

Kaul, Z.; Yaguchi, T.; Kaul, S. C.; Hirano, T.; Wadhwa, R.; Taira, K. Mortalin imaging in normal and cancer cells with quantum dot immuno-conjugates. *Cell Research*, 2003; 13:503–507.

Kamath, M. S.; Ahmed, S. S.; Dhanasekaran, M.; Santosh, S. W. Polycaprolactone scaffold engineered for sustained release of resveratrol: therapeutic enhancement in bone tissue engineering. *International Journal of Nanomedicine,* 2014;9:183-195.

Kawakami, S.; Sato, A.; Nishikawa, M.; Yamashita, F.; Hashida, M. Mannose receptor-mediated gene transfer into macrophages using novel mannosylated cationic liposomes. *Gene Therapy*, 2000; 7(4):292-299.

Kaye, S. B. New antimetabolites in cancer chemotherapy and their clinical impact. *British Journal of Cancer*, 1998; 78:(Suppl. 3):1.

Keyszer, G. M., Heer, A. H.; Kriegsmann, J.; Geiler, T.; Trabandt, A.; Keysser, M.; Gay, R. E.; Gay. S.; Comparative analysis of cathepsin L., cathepsin D., and collagenase messenger RNA expression in synovial tissues of patients with rheumatoid arthritis and osteoarthritis, by in situ hybridization. *Arthritis and Rheumatology*, 1995; 38:976.

Kitamura, K.; Takegami, S.; Tanaka, R.; Omran, A. A.; Kitade, T. Effect of long-chain Fatty acids on the binding of triflupromazine to human serum albumin: a spectrophotometric study. *Scientia Pharmaceutica*, 2013;82(2): 233-245.

Kim, J. A. Targeted therapies for the treatment of cancer. *American Journal of Surgey,* 2003; 186:264–268.

Kirac, Y.; Bilen, S.; Duranay, M. Comparison of laboratory findings in patients with glomerulonephritis classified according to histopathologic diagnosis. *Minerva Medica*, 2014; 105(2):149-156.

Kratz, F.; Fichtner, I.; Beyer, U. Antitumor activity of acid labile transferrin and albumin doxorubicin conjugates in in vitro and in vivo human tumour xenograft model. *European Journal of Cancer*, 1997; 33: 175–175.

Kratz, F. A clinical update of using albumin as a drug vehicle - a commentary. *Journal of Controlled Release*, 2014; 190:331-336.

Kratz, F. Albumin as a drug carrier: Design of prodrugs, drug conjugates and nanoparticles. *Journal of Controlled Release*, 2008; 132:171-183.

Kratz, F.; Beyer, U. Serum proteins as drug carriers of anticancer agents: a review. *Drug Delivery*, 1998; 5:281-299.

Kreuter J., Hekmatara T., Dreis S., et al. Covalent attachment of apolipoprotein AI and apolipoprotein B-100 to albumin nanoparticles enables drug transport to brain. *Journal of Controlled Release*, 2007; 118: 54–58.

Kundu, A., Nandi S., Das P., Nandi A. K. Fluorescent Graphene Oxide via Polymer Grafting: An Efficient Nanocarrier for Both Hydrophilic and Hydrophobic Drugs. *ACS Applied Material and Interfaces,* 2015; 7(6): 3512-3523.

Koh, S., Means, G. Characterization of a small apolar anion binding site of human serum albumin. *Archives of Biochemistry and Biophysics*, 1979; 192:73-79.

Levick, J. R. Permeability of rheumatoid and normal human synovium to specific plasma proteins. *Arthritis and Rheumatology*, 1981; 24: 1550-1560.

Lewis Phillips G. D., Li G., Dugger D. L., Crocker L. M., Parsons K. L. Targeting HER2-positive breast cancer with trastuzumab-DM1, an antibody-cytotoxic drug conjugate. *Cancer Research*, 2008; 68: 9280–9290.

Li, R.; Zheng, K.; Hu, P.; Chen, Z.; Zhou, S.; Chen, J.; Yuan, C.; Chen, S.; Zheng, W.; Ma, E.; Zhang, F.; Xue, J.; Chen, X.; Huang, M. A novel tumor targeting drug carrier for optical imaging and therapy. *Theranostics,* 2014; 4(6):642-659.

Li, X.; Wang, G.; Chen, D.; Lu,Y. β-Carotene and astaxanthin with human and bovine serum albumins. *Food Chemistry*, 2015; 179: 213–221.

Liberatore, M.; Clemente, M.; Iurilli, A. P.; Zorzin, L.; Marini, M.; Di Rocco, E.; Colella, A. C. Scintigraphic evaluation of disease activity in rheumatoid arthritis: a comparison of technetium-99m human non-specific immunoglobulins, leucocytes and albumin nanocolloids. *European Journal of Nuclear Medicine,* 1992. 19: 853-857.

Liu, M.; Huang, Y.; Hu, L.; Liu, G.; Hu, X.; Liu, D.; Yang, X. Selective delivery of interleukine-1 receptor antagonist to inflamed joint by albumin fusion. *BMC Biotechnology*, 2012;12: 68.

Llewelyn, C. A.; Hewitt, P. E.; Knight, R. S.; Amar, K.; Cousens, S.; Mackenzie, J.; Will, R. G. Possible transmission of variant Creutzfeldt-Jakob disease by blood transfusion. *Lancet,* 2004; 363:417-421.

Loganathan, R.; Ramakrishnan, S.; Suresh, E.; Palaniandavar, M.; Riyasdeen, A.; Akbarsha, M. A.; Mixed ligand μ-phenoxo-bridged dinuclear copper(II) complexes with diimine co-ligands: efficient chemical nuclease and protease activities and cytotoxicity. *Dalton Transactions,* 2014: 43(16): 6177-6194.

Lou, J.; Hu, W.; Tian, R.; Zhang, H.; Jia, Y.; Zhang, J.; Zhang, L. Optimization and evaluation of a thermoresponsive ophthalmic in situ gel containing curcumin-loaded albumin nanoparticles. *International Journal of Nanomedicine,* 2014; 21;9:2517-2525.

Lu, W.; Luo, H.; Zhu, Z.; Wu Y., Luo J., Wang, H. Preparation and the biopharmaceutical evaluation for the metered dose transdermal spray of dexketoprofen. *Journal of Drug Delivery*, 2014; 2014:697434.

Lu, Y.; Wang, Z. H.; Li, T.; McNally, H.; Park, K.; Sturek, M. Development and evaluation of transferrin-stabilized paclitaxel nanocrystal formulation. *Journal of Controled Released*, 2014; 28:176:76-85.

Miller, G. Colloid gold nanoparticles deliver cancer-fighting drugs. *Pharmacology Technology*, 2003; 7:17–19.

Maeda, H.; Wu, J.; Sawa, T.; Matsumura, Y.; Hori, K. Tumor vascular permeability and the EPR effect in macromolecular therapeutics: a review, *J. Controlled Release*, 2000;65: 271-284.

Mansour, A. M.; Drevs, J.; Esser, N.; Hamada, F. M.; Badary, O. A.; Unger, C.; Fichtner, I.; Kratz, F. A new approach for the treatment of malignant melanoma: enhanced antitumor efficacy of an albumin-binding doxorubicin prodrug that is cleaved by matrix metalloproteinase 2. *Cancer Research,* 2003; 63: 4062–4066.

Martin L. Human albumin solutions in the critical patient. *IVECCS Proceedings,* 2004; 274-278.

Matsumura, Y.; Maeda, H. A new concept for macromolecular therapeutics in cancer chemotherapy: mechanism of tumoritropic accumulation of proteins and the antitumor agent smancs. *Cancer Research*, 1986; 46: 6387-6392.

Meden, H.; Kuhn W. Overexpression of the oncogene c-erbB-2 (HER2/neu) in ovarian cancer: a new prognostic factor. *European Journal of Obstetrics and Gynecology,* 1997; 71: 173–179.

Mendez, C. M.; McClain, C. J.; Marsano, L. S. Albumin therapy in clinical practice. *Nutrition in Clinical Practice*, 2005; 20:314-320.

Merlot, A. M.; Kalinowski, D. S.; Richardson, D. R. Unraveling the mysteries of serum albumin-more than just a serum protein. *Frontier in Physiology,* 2014; 5:299.

Mitzner S. R., Klammt S., Peszynski P., Hickstein H., Korten G., Stange J., Schmidt R. Improvement of multiple organ functions in hepatorenal syndrome during albumin dialysis with the molecular adsorbent recirculating system. *Therapeutic Apheresis and Dialysis*, 2001; 5(5):417-22.

Moreland, L. W.; Baumgartner, S. W.; Schiff; M. H., Tindall E. A., Fleischmann R. M., Weaver, A. L.; Ettlinger, R. E.; Cohen, S.; Koopman, W. J.; Mohler, K.; Widmer, M. B.; Blosch, C. M. Treatment of rheumatoid arthritis with a recombinant human tumor necrosis factor receptor (p75)-Fc fusion protein. *New England Journal of Medicine,* 1997; 337: 141-147.

Mukherjee, B.; Karmakar, S. D.; Hossain, C. M.; Bhattacharya, S. Peptides, proteins and peptide/protein-polymer conjugates as drug delivery system. *Protein and Peptide Letters*, 2014; 21(11):1121-1128.

Müller-Ladner, U.; Gay, R. E.; Gay, S. Molecular biology of cartilage and bone destruction. *Current Opinion in Rheumatology*, 1998; 10: 212.

Ni, Z.; Yuan, Y.; Wang, Q.; Cao, L.; Che, X.; Zhang, M. Time-averaged albumin predicts the long-term prognosis of IgA nephropathy patients who achieved remission. *Journal of Translational Medicine,* 2014; 12(1): 194.

Niwa, Y.; Iio, A.; Niwa, G.; Sakane, T.; Tsunematsu, T.; Kanoh, T. Serum albumin metabolism in rheumatic diseases: relationship to corticosteroids and peptic ulcer. *Journal of Clinical Immunology*, 1990; 31: 11-16.

Nilvebrant, J.; Åstrand, M.; Georgieva-Kotseva, M.; Björnmalm, M.; Löfblom, J.; Hober, S. Engineering of bispecific affinity proteins with high affinity for ERBB2 and adaptable binding to albumin. *PLoS One,* 2014; 9(8):e103094.

Nicholson, J. P.; Wolmarans M. R.; Park, G. R. The role of albumin in critical illness. *British Journal of Anesthesia,* 2000; 599-610.

Neuzil, J.; Stocker, R. Free and albumin-bound bilirubin are efficient co-antioxidants for alpha-tocopherol, inhibiting plasma and low density lipoprotein lipid peroxidation. *Journal of Biological Chemistry*, 1994; 269:16712-9.

Nohynek, G. J.; Lademann, J.; Ribaud, C.; Roberts, M. S. Grey goo on the skin? Nanotechnology, cosmetic and sunscreen safety. *Critical Review of Toxicology,* 2007; 37:251–277.

Noguchi, Y.; Wu, J.; Duncan, R.; Strohalm, J.; Ulbrich, K.; Akaike, T.; Maeda, H.; Early phase tumor accumulation of macromolecules: a great difference in clearance rate between tumor and normal tissues. *Japanese Journal of Cancer Research*, 1998; 89:307-314.

O'Dell, J. R. Methotrexate use in rheumatoid arthritis. *Rheumatic Disease Clinics of North America,* 1997; 23:779.

Oettinger, C. W.; D'Souza, M. J. Microencapsulated drug delivery: a new approach to pro-inflammatory cytokine inhibition. *Journal of Microencapsulation,* 2012;29: 455-462.

Paciotti, G. F.; Myer, L.; Weinreich, D.; Goia, D.; Pavel, N.; McLaughlin, R. E.; Tamarkin, L. Colloidal gold: A novel nanoparticle vector for tumour directed drug delivery. *Drug Delivery*, 2004; 11:169–183.

Pandey, R.; Ahmad, Z.; Sharma, S.; Khuller, G. K. Nano-encapsulation of azole antifungals: Potential applications to improve oral drug delivery. *International Journal of Pharmacology*, 2005; 301:268–276.

Pap, T.; Müller-Ladner, U.; Gay, R. E.; Gay, S. Fibroblast biology: role of synovial fibroblasts in the pathogenesis of rheumatoid arthritis. *Arthritis Research,* 2000; 2:361.

Papazisis K. T., Habeshaw T., Miles D. W. Safety and efficacy of the combination of trastuzumab with docetaxel for HER2-positive women with advanced breast cancer. A review of the existing clinical trials and results of the expanded access programme in the UK. *International Journal of Clinical Practice,* 2004; 58: 581–586.

Paulos, C. M.; Turk, M. J.; Breur, G. J.; Low, P. S. Folate receptor-mediated targeting of therapeutic and imaging agents to activated macrophages in rheumatoid arthritis. *Advanced Drug Delivery Review*, 2004; 56: 1205-1217.

Peters, T. All About Albumin: Biochemistry, Genetics and Medical Applications. San Diego, CA: Academic Press Limited. 1996.

Peters, T. Serum albumin. *Advanced Protein Chemistry*, 1985; 37: 161-245.

Petitpas, I.; Bhattacharya, A.; Twine, S.; East, M.; Curry, S. Crystal structure analysis of warfarin binding to human serum albumin. *Journal of Biological Chemistry,* 2001; 276:22804-22809.

Peralta, D. V.; He, J.; Wheeler, D. A.; Zhang, J. Z.; Tarr, M. A. Encapsulating gold nanomaterials into size-controlled human serum albumin nanoparticles for cancer therapy platforms. *Journal of Microencapsulation,* 2014; 31(8):824-831.

Popov, A. P.; Lademann, J.; Priezzhev, A. V.; Myllylä, R. Effect of size of TiO2 nanoparticles embedded into stratum corneum on ultraviolet-A and ultraviolet-B sun-blocking properties of the skin. *Journal of Biomedical Optics,* 2005; 10:1–9.

Rahnama E., Mahmoodian-Moghaddam M., Khorsand-Ahmadi S., Saberi M. R., Chamani J. Binding site identification of metformin to human serum albumin and glycated human serum albumin by spectroscopic and molecular modeling techniques: a comparison study. *Journal of Biomolecular Structure and Dynamics,* 2015; 33(3):513-33.

Reed, R. G. A Trail of Research from Lipoic Acid to α-Keto Acid Dehydrogenase Complexes *Journal of Biological Chemistry*, 1977; 252: 7483-7487.

Richieri G., Anel A., Kleinfeld A. Interaction of long-chain fatty acids and albumin: determination of free fatty acid levels using the fluorescent probe ADIFAB. *Biochemistry,* 1993; 2:7574-7580.

Ríhová, B.; Jelínková, M.; Strohalm, J.; Subr, V.; Plocová, D., Hovorka, O.; Novák, M.; Plundrová, D.; Germano, Y.; Ulbrich, K. Polymeric drugs based on conjugates of synthetic and natural macromolecules. II. Anti-cancer activity of antibody or (Fab')(2)-targeted conjugates and combined therapy with immunomodulators. *Journal of Controled Release,* 2000; 64(1-3):241-261.

Rink, T.; Heuser, T.; Fitz, H.; Schroth, H. J.; Weller, E.; Zippel, H. H. Lymphoscintigraphic sentinel node imaging and gamma probe detection in breast cancer with Tc-99m nanocolloidal albumin: results of an optimized protocol. *Clinical Nuclear Medicine,* 2001; 26: 293-298.

Rollett, A.; Reiter, T.; Nogueira, P.; Cardinale, M.; Loureiro, A.; Gomes, A.; Cavaco-Paulo, A.; Moreira, A.; Carmo, A. M.; Guebitz, G. M. Folic acid-functionalized human serum albumin nanocapsules for targeted drug delivery to chronically activated macrophages. *International Journal of Pharmaceutics,* 2012; 427: 460-466.

Sajid Ali, M.; Al-Lohedan, H. A.; Atta, A. M.; Ezzat, A. O.; Al-Hussain, S. A. A. Interaction of human serum albumin with silver nanoparticles functionalized with polyvinylthiol. *Journal of Molecular Liquids,* 2015; 204: 248–254.

Schmid, B.; Chung, D. E.; Warnecke, A.; Fichtner, I.; Kratz, F. Albumin-binding prodrugs of camptothecin and doxorubicin with an Ala-Leu-Ala-Leu-linker that are cleaved by cathepsin B: synthesis and antitumor efficacy. *Bioconjugate Chemistry*, 2007; 18: 702–716.

Shahabadi, N.; Hadidi, S.; Feizi, F. Study on the interaction of antiviral drug 'Tenofovir' with human serum albumin by spectral and molecular modeling methods. *Spectrochim Acta A Molecular Biomolecular Spectroscopy*, 2015; 138:169-175.

Siwale, R. C.; Oettinger, C. W.; Pai, S. B.; Addo, R.; Uddin, N.; Siddig, A.; D'Souza, M. J. Formulation and characterization of catalase in albumin microspheres. *Journal of Microencapsulation*, 26: 411-419.

Shahzad, Y.; Khan, I. U.; Hussain, T.; Alamgeer Serra, C. A.; Rizvi, S. A.; Gerber, M.; du Plessis, J. Bioactive albumin-based carriers for tumour chemotherapy. *Currunt Cancer Drug Targets*, 2014; 14(8):752-763.

Shenoy, D. B.; Amiji, M. M. Poly(ethylene oxide)-modified poly(epsilon-caprolactone) nanoparticles for targeted delivery of tamoxifen in breast cancer. *International Journal of Pharmacy*, 2005; 293(1–2):261–270.

Schuitmaker, J. J.; Baas, P.; van Leengoed, H. L. L. M.; van der Meulen, F. W.; Star, W. M.; van Zandwijk, N. Photodynamic therapy: A promising new modality for the treatment of cancer. *Journal of Photochemistry and Photobiology B,* 1996; 34:3–12.

Simões, S.; Slepushkin, V.; Pires, P.; Gaspar, R.; Pedroso de Lima, M. C.; Düzgüneş, N. Human serum albumin enhances DNA transfection by lipoplexes and confers resistance to inhibition by serum. *Biochimica and Biophysica Acta,* 2000; 1463(2):459-469.

Sivertsen, A.; Tørfoss, V.; Isaksson, J.; Ausbacher, D.; Anderssen, T.; Brandsdal, B. O.; Havelkova, M.; Skjørholm, A. E.; Strøm, M. B. Anticancer potency of small linear and cyclic tetrapeptides and pharmacokinetic investigations of peptide binding to human serum albumin. *Journal of Peptide Science*, 2014; 20(4):279-291.

Schwartz, C. L. Late effects of treatment in long-term survivors of cancer. *Cancer Treatment Reviews*, 1995; 21:355–366.

Souza, N. C.; Simões, B. P.; Júnior, A. A.; Chiarello, P. G. Changes in intestinal permeability and nutritional status after cytotoxic therapy in patients with cancer. *Nutrition in Cancer*, 2014; 66(4):576-82.

Stehle, G.; Wunder, A.; Schrenk, H. H.; Hartung, G.; Heene, D. L.; Sinn, H. Albumin-based drug carriers: comparison between serum albumins of different species on pharmacokinetics and tumor uptake of the conjugate. *Anticancer Drugs*, 1999; 10(8):785-790.

Stehle, G.; Sinn, H.; Wunder, A.; Schrenk, H. H.; Stewart, J. C.; Hartung, G.; Maier-Borst, W.; Heene, D. L. Plasma protein (albumin) catabolism by the tumor itself-implications for tumor metabolism and the genesis of cachexia. *Critical* Reviews in *Oncology/Hematology,* 1997; 26: 77-100.

Steeland, S.; Puimège, L.; Vandenbroucke, R. E.; Van Hauwermeiren; F., Haustraete, J; Devoogdt, N; Hulpiau, P; Leroux-Roels, G; Laukens, D; Meuleman, P.; De Vos, M.; Libert, C. Generation and characterization of small single domain antibodies inhibiting human tumor necrosis factor receptor 1. *Journal of Biological Chemistry*, 2015; 290(7):4022-4037.

Steinhauser, I.; Spankuch, B.; Strebhardt, K.; Langer, K. Trastuzumab-modified nanoparticles: Optimisation of preparation and uptake in cancer cells. *Biomaterials,* 2006; 27:4975–4983.

Sun, S.; Tang, L.; Zhang, J.; Lv, F.; Wang, Z.; Wang, L.; Zhang, Q.; Zheng, C.; Qiu, L.; Jia, Z.; Lu, Y.; Liu, G.; Shao, Z.; Wang, B.; Hu, X. Cisplatin improves antitumor activity of weekly nab-paclitaxel in patients with metastatic breast cancer. *International Journal of Nanomedicine,* 2014; 19; 9:1443-1452.

Sugio, S.; Kashima, A.; Mochizuki, S.; Noda, M.; Kobayashi, K. Crystal structure of human serum albumin at 2.5 A resolution. *Protein Engineering,* 1999; 12:439-446.

Sudlow, G.; Birkett, D. J.; Wade, D. N. The characterization of two specific drug binding sites on human serum albumin. *Molecular Pharmacology,* 1975; 11(6):824-832.

Tormey, D. C.; Gray, R.; Taylor, S. G. Postoperative chemotherapy and chemohormonal therapy in women with node-positive breast cancer. *National Cancer Institute's Monography*, 1986; 1:75–80.

Tishler, M.; Lysyy, O.; Levy, O.; Volkov, O.; Golan, H. 99m Tc-albumin nanocolloid joint scintigraphy in rheumatoid arthritis patients who are in clinical remission - is remission real? *Clinical and Experimental Rheumatology,* 2010; 28: 360-364.

Tirsit Mogues, Junzhi Li, John Coburn, David J. Kuter. IgG antibodies against bovine serum albumin in humans—their prevalence and response to exposure to bovine serum albumin. *Journal of Immunological Methods*, 2005; 300:1–2, 1-11.

Totoribe K., Takamura N., Ogata K., Tokunaga J., Kawai K., Nakamura K., Ieiri I., Higuchi S. Influences of artificial heart-lung machine operation on the binding sites of albumin: possibility of an effective administration plan. *International Journal of Artificial Organs*, 2014; 37(2):99-108.

Turk, M. J.; Breur, G. J.; Widmer, W. R.; Paulos, C. M.; Xu, L. C.; Grote, L. A.; Low, P. S. Folate-targeted nanoparticles show efficacy in the treatment of inflammatory arthritis. *Arthritis and Rheumatology*, 2002; 63: 2671-2680.

Van der Heijden, J. W.; Oerlemans, R.; Dijkmans, B. A.; Qi ,H.; van der Laken, C. J.; Lems, W. F.; Jackman, A. L.; Kraan, M. C.; Tak, P. P.; Ratnam, M.; Jansen, G. Folate receptor beta as a potential delivery route for novel folate antagonists to macrophages in the synovial tissue of rheumatoid arthritis patients. *Arthritis and Rheumatology*, 2009; 60: 12-21.

Van der Zee, J. Heating the patient: A promising approach? *Annals Oncology,* 2002; 13:1173–1184.

Varshosaz, J.; Sadeghi-aliabadi, H.; Ghasemi, S.; Behdadfar, B. Use of magnetic folate *BioMed. Research International,* 2013; 2013:680712.

Voura, E. B.; Jaiswal, J. K.; Mattoussi, H.; Simon. S. M. Tracking metastatic tumor cell extravasation with quantum dot nanocrystals and fluorescence emission-scanning microscopy. *Nature Medicine*, 2004; 10:993–998.

Wacker, M. G.; Altinok, M.; Urfels, S.; Bauer, J. Nanoencapsulation of ultra-small superparamagnetic particles of iron oxide into human serum albumin nanoparticles. *Beilstein Journal of Nanotechnology*, 2014;5:2259-2266.

Wan, X.; Zheng, X.; Pang, X.; Zhang, Z.; Jing, T.; Xu W., Zhang Q. The potential use of lapatinib-loaded human serum albumin nanoparticles in the treatment of triple-negative breast cancer. *International Journal of Pharmacology,* 2015;484(1-2):16-28.

Wang, J.; Mongayt, D.; Torchilin, V. P. Polymeric micelles for delivery of poorly soluble drugs: Preparation and anticancer activity in vitro of paclitaxel incorporated into mixed micelles based on poly(ethylene glycol)-lipid conjugate and positively charged lipids. *Journal of Drug Targeting,* 2005; 13:73–80.

Wang, X. Y.; Koller, R.; Wirth, M.; Gabor, F. Lectin-Grafted PLGA Microcarriers Loaded with Fluorescent Model Drugs: Characteristics, Release Profiles, and Cytoadhesion Studies. *Scientia Pharmaceutica*, 2014; 82(1):193-205.

Warnecke, A.; Fichtner, I.; Sass, G.; Kratz, F. Synthesis, cleavage profile, and antitumor efficacy of an albumin-binding prodrug of methotrexate that is cleaved by plasmin and cathepsin B. *Archived Pharmacology*, (Weinheim) 2007; 340: 389–395.

Wee, K. W.; Kang, G. Y.; Park, J.; Kang, J. Y.; Yoon, D. S.; Park, J. H.; Kim, T. S. Novel electrical detection of label-free disease marker proteins using piezoresistive self-sensing micro-cantilevers. *Biosensors and Bioelectronics,* 2005; 20:1932–1938.

Weiner, N.; Lieb, L.; Niemiec, S.; Ramachandran, C.; Hu, Z.; Egbaria, K. Liposomes: A novel topical delivery system for pharmaceutical and cosmetic applications. *Journal of Drug Targeting*, 1994; 2:405–410.

Wust, P.; Hildebrandt, B.; Sreenivasa, G. Hyperthermia in combined treatment of cancer. *Lancet Oncology*, 2002; 3:487–497.

Xu, X.; Qian,Y.; Wu, P.; Zhang, H.; Ca, Ch. Probing the anticancer-drug-binding-induced microenvironment alterations in subdomain IIA of human serum albumin. *Journal of Colloid and Interface Science*, 445; 2015 102-111.

Xu, B.; Yu, H. B.; Hui, W.; He, J. L.; Wei, L. L.; Wang, Z. Clinical features and risk factors of acute hepatitis E with severe jaundice. *World Journal of Gastroenterology,* 2012; 18(48):7279-7284.

Xu, X.; Qian, Y.; Wu, P.; Zhang, H.; Cai, C. Probing the anticancer-drug-binding-induced microenvironment alterations in subdomain IIA of human serum albumin. *Journal of Colloid and Interface Science,* 2015; 445:102-111.

Yeggoni, D. P.; Rachamallu, A.; Kallubai, M.; Subramanyam, R. Cytotoxicity and comparative binding mechanism of piperine with human serum albumin and α-1-acid glycoprotein. *Journal of Biomolecular Structure and Dynamic,* 2014:1-16.

Yoshikawa, T.; Uchida, T.; Naito, S.; Horino, A.; Taneichi, M.; Kato, H.; Komuro, K.; Nakano, Y.; Mori, M.; Nishinohara, S.; Chiba, J.; Kurata, T.; Tamura, S. Suppression of specific IgE antibody responses by liposome-conjugated ovalbumin in mice sensitized with ovalbumin via the respiratory tract. *International Archives of Allergy and Immunology,* 2000; 121(2):108-115.

Yu, Z.; Yu, M.; Zhou, Z.; Zhang, Z.; Du, B.; Xiong, Q. Novel flower-shaped albumin particles as controlled-release carriers for drugs to penetrate the round-window membrane. *International Journal of Nanomedicine*, 2014; 9:3193-3201.

Yuan, F.; Dellian, M.; Fukumura, D.; Leunig, M.; Berk, D. A.; Torchilin, V. P.; Jain, R. K. Vascular permeability in a human tumor xenograft: molecular size dependence and cutoff size. *Cancer Research*, 1995; 55: 3752–3756.

Yoo, H. S.; Oh, J. E.; Lee, K. H.; Park, T. G. Biodegradable nanoparticles containing doxorubicin-PLGA conjugate for sustained release. *Pharmacology Research*, 1999; 16:1114–1118.

Yeggoni, D. P.; Rachamallu, A.; Kallubai, M.; Subramanyam, R. Cytotoxicity and comparative binding mechanism of piperine with human serum albumin and α-1-acid glycoprotein. *Journal of Biomolecular Structure and Dynamic,* 2014:1-16.

Yoshikawa, T.; Uchida, T.; Naito, S.; Horino, A.; Taneichi, M.; Kato, H.; Komuro, K.; Nakano, Y.; Mori, M.; Nishinohara, S.; Chiba, J.; Kurata, T.; Tamura, S. Suppression of specific IgE antibody responses by liposome-conjugated ovalbumin in mice sensitized with ovalbumin via the respiratory tract. *International Archives* of *Allergy* and *Immunology,* 2000; 121(2):108-115.

Yu, Z.; Yu, M.; Zhou, Z.; Zhang, Z.; Du, B.; Xiong, Q. Novel flower-shaped albumin particles as controlled-release carriers for drugs to penetrate the round-window membrane. *International Journal of Nanomedicine,* 2014; 9:3193-3201.

Zaloga, J.; Janko, C.; Nowak, J.; Matuszak, J.; Knaup, S.; Eberbeck, D.; Tietze, R.; Unterweger, H.; Friedrich, R. P.; Duerr, S.; Heimke-Brinck, R.; Baum, E.; Cicha, I.; Dörje, F.; Odenbach, S.; Lyer, S.; Lee, G.; Alexiou, C. Development of a lauric acid/albumin hybrid iron oxide nanoparticle system with improved biocompatibility. *International Journal of Nanomedicine,* 2014;9:4847-4866.

Zhao, Z.; Wang, Y.; Han, J.; Wang, K.; Yang, D.; Yang, Y.; Du, Q.; Song, Y.; Yin, X. Self-assembled micelles of amphiphilic poly(L-phenylalanine)-b-poly(L-serine) polypeptides for tumor-targeted delivery. *International Journal of Nanomedicine,* 2014;9:5849-5862.

Zharov, V. P.; Galitovskaya, E. N.; Johnson, C.; Kelly, T. Synergistic enhancement of selective nanophotothermolysis with gold nanoclusters: Potential cancer therapy. *Laser in Surgery and Medicine*, 2005; 39:219–226.

Zheng, Y. R.; Suntharalingam, K.; Johnstone, T. C.; Yoo, H.; Lin, W.; Brooks, J. G.; Lippard, S. J. Pt(IV) prodrugs designed to bind non-covalently to human serum albumin for drug delivery. *Journal of American and Chemical Society,* 2014; 136(24):8790-8798.

In: Human Serum Albumin (HSA)
Editor: Travis Stokes

ISBN: 978-1-63482-963-2

Chapter 3

RECOMBINANT HUMAN SERUM ALBUMIN FUSION PROTEINS: NOVEL APPLICATIONS AND CHALLENGES

Daoyuan Dong[2], Zhijun Li[2], Xun Li[3*] and Zhiyu Li[1*]

[1]Department of Pharmaceutical Sciences, Philadelphia College of Pharmacy
[2]Department of Chemistry and Biochemistry,
Misher College of Arts and Sciences, University of the Sciences, Philadelphia, PA, US
[3]Department of Medicinal Chemistry, School of Pharmaceutical Science,
Shandong University, Ji'nan, China

ABSTRACT

Human serum albumin (HSA) is the most abundant plasma protein. It has been extensively studied and applied in medicine and pharmaceutics. Taking advantage of the recombinant DNA technology, fusion proteins with novel functions have been generated. The recent approval of the albumin-GLP-1 fusion protein drug, Tanzeum, will encourage researchers to explore many more advantages of HSA for therapeutic applications. Generally, these albumin fusion proteins possess long serum half-life, good solubility, and versatile ligand binding ability. In addition to being used as drugs, albumin fusion proteins can serve as carriers for drug delivery to improve drug efficacy and safety. Herein, novel investigations on the applications of albumin fusion proteins will be discussed, with focus on tumor targeting and intracellular delivery. These approaches carry the potential to co-deliver multiple drugs to the same cellular target for synergistic therapeutic effects.

1. INTRODUCTION

Human serum albumin (HSA) is the most abundant protein in plasma and has a long history in clinical practice for the treatments of hypovolemia, shock, burns, surgical blood loss, trauma, and hemorrhage, dating back to World War II [1]. Ideal biological and

[*] Corresponding authors: zh.li@usciences.edu (Z.L.), tjulx2004@sdu.edu.cn (X.L.)

physicochemical properties of albumin, such as long serum half-life, low immunogenicity, extraordinary ligand binding ability, good solubility, and thermal stability make albumin a good pharmaceutical carrier for drug delivery. Many papers have thoroughly reviewed the applications of albumins in drug discovery [2–4]. One of the essential applications of albumin in biopharmaceutics is to construct recombinant HSA fusion proteins to improve the pharmacokinetic profiles of recombinant therapeutic proteins [5].

Therapeutic proteins/peptides with low molecular weight are rapidly eliminated in plasma. Extending the hydrodynamic size or increasing the molecular weight of therapeutics by nano/micro particle encapsulation [6], PEGylation [7], glycosylation [8], and fusion proteins [9] are common strategies to develop long-lasting protein therapeutics. With the development of recombinant DNA technology, researchers are able to design a variety of fusion proteins with novel functions that do not exist in nature. IgG antibody Fc fragment fusion therapeutic proteins, such as Enbrel (etanercept), Orencia (abatacept), and Nplate (romiplostim), have been successfully developed and routinely prescribed in clinical practice. The first HSA fusion protein drug, Tanzeum (albiglutide), was also approved by the FDA [10]. However, the application of albumin fusion proteins is not only limited on extending the plasma half-life of therapeutic proteins. Albumin fusion proteins still possess the extraordinary ligand binding capability of wild type albumin. They can carry and deliver more than one type of drug to target multiple intracellular pathways of tumor cells.

Albumin is capable of transporting saturated long-chain fatty acids (LCFAs) with multiple binding sites and high affinity [11]. Meanwhile, albumin can be genetically fused with protein/peptide to target the surface biomarkers or essential intracellular pathways of tumor cells. Complexing albumin fusion proteins with fatty acid (FA)-modified drugs by non-covalent binding will allow us to deliver long-lasting drugs into tumor cells. Interestingly, tumor cells tend to take albumins and fatty acids to maintain altered metabolism [12]. Therefore, in addition to the conventional half-life extension benefit, albumin fusion proteins can function like multifunctional drugs as well as drug carriers for synergistic tumor cytotoxicities.

2. Fusion Proteins

Recombinant fusion protein can be generated naturally by chromosomal translocation or inversion in human cells, such as oncogenic BCR-ABL kinase possessed by 95% chronic myelogenous leukemia (CML) [13], or constructed by genetically combining two individual proteins or peptides into one protein or polypeptide through recombinant DNA technology for novel functions. His-tag, FLAG-tag, GST fusion, and GFP fusion have been commonly practiced in protein cloning, affinity purification, and histochemistry imaging.

Most therapeutic fusion proteins fit in the triple T (T3) paradigm (Time, Toxin and Targeting) [9]. To realize each T, there are pools of modules to choose from. For Time, common strategies include fusing small proteins with IgG Fc fragment [14], transferrin [15] or albumin [16]. For Toxin, proteins are usually fused with peptide toxins [17], enzymes [18] or cytokines [19] to produce a cytotoxic effect. Finally, for Targeting, proteins are combined with modules such as GILT (glycosylaton independent lysosomal targeting) [20] or apoA1 (apolipoprotein A1) [21] to gain novel targeting functions. Most fusion proteins are the combinations of two or three of these modules to maintain multi-functionality.

Fusing a tumor-targeting peptide with a cytotoxic protein/peptide is a well-established approach of developing antineoplastics. Ontack (denileukin diftitox) is a fusion protein of IL-2 and diphtheria toxin that was approved for cutaneous T-cell lymphoma [22]. In addition to targeted delivery of a toxin, the main purpose of construction of a fusion protein is to extend the plasma half-life of a protein drug by simply adding molecular weight.

Generally, fusion proteins are easy to construct, express, purify, and formulate. IgG Fc fragment, HSA, transferrin, apolipoprotein A1, and artificial XTEN have been studied as fusion protein modules. IgG Fc fragment fusion proteins, such as Enbrel, Orencia, and Nplate, have been approved by the FDA. IgG Fc fragment possesses binding sites to neonatal Fc receptor (FcRn).

It has been found that binding to FcRn is the key mechanism to recycle IgG from vesical endothelial degradation and prolong the plasma half-life [23]. Therefore, FcRn binding and recycling is another mechanism to make long-lasting protein therapeutics in addition to adding molecular weight. Fc fusion proteins may form dimers through the Fc fragment and cause heterogeneity during formulation. Additionally, glycosylation, antibody-dependent cell-mediated cytotoxicity (ADCC), and complement-dependent cytotoxicity (CDC) are also associated with the Fc fragment. It can affect the therapeutic efficacy depending on whether these functions are necessary [24].

3. Albumin Fusion Proteins

HSA is an abundant protein in human plasma (35~50 g/L) with 585 amino acids and a molecular weight about 67 kDa. It is known to bind, solubilize, and transport a variety of endogenous and exogenous compounds in blood, such as fatty acids, metal ions, therapeutic drugs and metabolites [2]. After being synthesized in liver (10~15 g/day), 30~40% of albumins are kept in plasma, and the rest will be located in extravascular tissues [11]. Recent studies have confirmed the interaction between albumin and FcRn [25]. It reveals why albumin has an approximate plasma half-life of 19 days. Biochemically, albumin is a non-glycosylated protein composed of three homologous domains with dominant α-helical foldings. Seventeen pairs of disulfide bonds hold intra-domain α-helices [11]. This unique structure contributes significantly to the thermal stability, but does not restrain the flexibility of each domain. The unpaired free cysteine at residue 34 is believed to act as an antioxidant to protect the albumin-bound substances from oxidation or a scavenger to reduce reactive oxygen and nitrogen species [26]. It is also the site to form an intermolecular disulfide bond and make an albumin dimer [27].

These unique biological and biophysical properties make albumin an ideal pharmaceutical material and drug carrier. One commonly practiced approach is to genetically fuse HSA with therapeutic proteins. Gaining molecular weight by fusing with albumin limits the kidney filtration of therapeutic proteins and then prolongs the plasma retention. Meanwhile, FcRn binding may also contribute to the recycling of albumin fusion protein in vesicular endothelial cells [28]. Although *in vitro* studies have shown that N- or C-terminal single-chain variable fragment (scFv) or short peptide fusion only slightly affected the affinity to FcRn [29], more detailed *in vivo* and animal works are expected to assess the role of FcRn on the plasma half-life of albumin fusion proteins.

Construction of an albumin fusion protein is straightforward through recombinant DNA technology. Production can be easily scaled up in a well-established yeast expression system (*P pastoris or S cerevisiae*). Wild-type HSA is non-glycosylated, therefore no post-translational concern needs to be addressed [27]. However, if the fusion partner requires sophisticated post-translational modification, other expression systems, such as mammalian cells (CHO or HEK293) should be considered. The purification procedure is also easy and economical through either affinity purification or few conventional chromatography steps.

Despite easy construction, it requires careful design and experience to obtain a fusion protein with desired therapeutic efficacy. For example, based on the structure/function of a therapeutic protein, a proper fusion orientation (N- or C-terminal fusion) and linkers (flexible, rigid, or cleavable linkers and the length of a linker) need to be optimized and selected. Due to the different biochemical and biophysical properties of albumin and a therapeutic protein, optimal expression system, purification strategies, and formulation are crucial for the stability and activity of fusion proteins. Although many clinical trials have indicated the low immunogenicity of albumin fusion proteins, unfolded or partially unfolded proteins generated during formulation and storage may not only affect efficacy but also be recognized as immunogens.

In addition, the fusion protein orientation may affect the protein activity. In one study, both N and C terminal interleukin-28B-albumin fusion proteins (HSA-IL28B) were obtained, while only the IL28B fused to the N-terminus of albumin showed *in vitro* antiviral activity [30]. Another report showed the N-terminal albumin fusion interferon-α2b (IFN-α2b-HSA) exhibited better homogeneity and stability than its C-terminal counterpart, at the expense of a slightly decreased antiviral activity [31]. The linker of a fusion protein is also needed to be selected for an optimal activity. The same group also investigated the effects of different linker sequences on the stability and activity of N-terminal fusion IFN-α2b-HSA [32]. Fusion proteins with three types of linkers, rigid, flexible, and helical linkers, have been constructed. It was found that inserting a linker between the two fusion components was able to increase the stability. All constructions with linkers showed improved antiviral activity compared with the original fusion protein with no linker. Furthermore, separating the steric hindrance between bulky albumin and fusion target may facilitate the flexibility and activity of the fusion protein. A paper has given a thorough review on fusion protein linker design [33]. Usually, multiple constructions with various orientations, linkers and even tandem repeats of the fusion target should be constructed to screen an optimal structure. Additionally, molecular modeling and biophysical characterization could be employed to facilitate the fusion protein design and screen.

A recent review has comprehensively summarized the development of albumin fusion proteins [16]. Here, an update of albumin fusion proteins is presented (Table 1). These fusion proteins bind to versatile receptors on cellular surface and have been applied to treat cancer, autoimmune disease, infectious disease, thrombosis, or diabetes. Unlike the fruitful IgG Fc fusion proteins, few albumin fusion proteins have been evaluated in the late stage of clinical trials, and among which, only albumin-GLP-1 fusion protein drug, Tanzeum, was approved by the FDA. Table 2 lists albumin fusion proteins in clinical trials. Nevertheless, albiglutide is a breakthrough of utilizing albumin fusion strategy. It definitely eases the immunogenicity concerns of long-term use of albumin fusion proteins. Most importantly, more research will be conducted to develop novel applications of albumin fusion proteins.

Table 1. Summary of Recombinant Albumin Fusion Proteins

Albumin Fusion Partner	N- or C- fuse	Linker	Expressing System	Application	Reference
Interferon-α2b (INF-α2b)	N- or C-	Various	P pastoris	Hepatitis C	[32]
Glucagon-like peptide (GLP-1)	N-	$(GGGGS)_2GGGAS$	CHO-K1	Reduce blood glucose	[90]
Exendin-4	N-	GGGGS	P pastoris	Improve glucose tolerance	[91]
Granulocyte colony stimulating factor (G-CSF)	C-	N.A.	S cerevisiae	Promote neutrophils growth	[92]
Interleukin-2 (IL-2)	N-	No linker	P pastoris	Mediate anti-tumor cellular immune responses	[93]
Interleukin-28B (IL-28B)	N-	No linker	P pastoris	Anti-HCV	[30]
Insulin	N-	Dodecapep-tide	CHO	Long-acting insulin analog	[94]
Human Growth factor (hGH)	C-	N.A.	S cerevisiae	Stimulate IGF-1	[95]
Soluble EphB4	N-	N.A.	CHO	Regulate vessel maturation	[96]
Butyrylcholinesterase (BChE)	N-	G_6S	BHK	Prevent oranophosphorus compound toxicity	[97]
B-type natriuretic peptide (BNP)	N-	No linker	293F	Reduce blood pressure	[98]
Erythropoietin (EPO)	C-	$(GGSGG)_4$	CHO	Increase the hematocrit	[99]
Barbourin	N-	G_6S	P pastoris	Inhibit platelet aggregation	[100]
Hirudin	N-	N.A.	P pastoris	Thrombin inhibitor	[101]
FVIIa	N-	31 aa G-S linker	HEK293	Inhibit hemophilia	[102]
FIX	N-	SVSQTSKLTRAETVFPDVDGS	CHO	Promote coagulation	[48]
Infestin-4	C-	$(GGS)_3$	E coli	Inhibit FXIIa	[103]
Thioredoxin	C-	$(GGGGS)_2$	P pastoris	Anti-inflammatory	[104]
Thymosin-α1 (T-α1)	C-	$(GGGGS)_2$	P pastoris	Stimulate immune response	[105]
TNF ligands	C-	Dodecapep-tide	HEK293	Induce apoptosis	[106]
Soluble human tumor necrosis factor receptors (shTNFR)	N- or C-	$(GGGGS)_2$	P pastoris	Neutralizing TNF-α	[107]
scFv antibody fragments	N- and/or C-	$(GGS)_4GG$	S cerevisiae	Increase bioavailability	[108]
Bispecific Ab: taFv, Db, scDb	N- and/or C-	Various	E coli	Targeting CEA and CD3	[109]

Table 2. Summary of Recombinant Albumin Fusion Proteins in Clinical Trials

Albumin Fusion Partner	Name	Clinical trial status	Therapeutic indication	Reference
GLP-1 dimer	Albiglutide/albugon	Approved (2014) Tanzeum	Type 2 diabetes	NCT00849017, NCT01098539, NCT01733758, NCT01128894, NCT00839527, NCT00838903, NCT00849056, NCT00838916
rhG-CSF	Albugranin/Neugranin/ Balugrastim	III completed, withdrawn by Teva due to marketing strategy	Chemotherapy-induced Neutropenia	NCT01126190, NCT00837265
	GW003	I (Jiangsu T-Mab Biopharma)		NCT02156388
FIX	rIX-FP	III	Hemophilia B	NCT02053792
hGH	Albutropin/ TV-1106	II	Growth hormone deficiency	NCT02092077
BNP(Cardeva)	AlbuBNP,	I/II	Congestive heart failure	(98)
FVIIa	rVIIa-FP	I	Hemophilia A	NCT01542619
sEphB4	sEphB4-HSA	I	Advance solid tumors	NCT01642342
IL-2	Albuleukin	I	Cancer	(110)
INF-α2b	Albuferon	Discontinued due to serious pulmonary adverse events	Hepatitis C	NCT00724776, NCT00964665, NCT00489385, NCT00759200

3.1. Glucagon-Like Peptide 1 (GLP-1)

GLP-1 is secreted from intestinal L-cells in response to nutrient ingestion. The active GLP-1 (residues 7~36) acts on GLP-1 receptors of pancreatic beta cells to stimulate glucose-dependent insulin secretion and inhibit glucagon secretion [34]. It also delays gastric emptying and promotes satiety. GLP-1 has a very short half-life (< 2 minutes) due to the dipeptidyl peptidase-4 (DPP IV) cleavage at the residue 8 alanine. Byetta (exenatide) is a synthetic DPP IV-resistant GLP-1 homologue with a plasma half-life of 2.4 hours [35]. The once weekly dosage form of exenatide, Bydureon (exenatide LAR), is an poly-(D,L-lactide-co-glycolide) (PLG) microsphere with exendin-4 encapsulation [36]. Albiglutide, the first albumin fusion protein drug, is another once weekly GLP-1 analogue [37].

Two tandem copies of GLP-1 analogue are fused to the N-terminus of albumin to obtain an Albiglutide. To avoid peptidase cleavage, the residue 8 alanine is replaced by glycine in GLP-1. As discussed above, the steric hindrance between bulky albumin and the receptor binding region of GLP-1 is prone to interfere with the function of GLP-1 peptide. Albiglutide only showed 1% of exenatide activity characterized by cAMP generation *in vitro*. However, the very long half-life and high administration dose is able to compensate the low *in vitro* activity of this fusion protein.

3.2. Interferon α (INF α)

INF α belongs to type I interferon. It is secreted by virus-infected leukocytes and has antiviral, antiproliferative, and immunostimulatory bioactivities [38]. At present, recombinant INF α2a and INF α2b have been approved to treat hepatitis B and C, and certain cancers. In addition, two types of PEGlyated INF α long-lasting formulation are available in market. Pegasys is a mixture of pegylated INF α2a (PEG-INF α2a) isomers with approximate 65 hours of plasma half-life and 3~7 % of native INF α2a bioactivity [39]. Pegintron is a mixture of pegylated INF α2b (PEG-INF α2b) isomers with approximate 40 hours of plasma half-life and 27% of INF α2b bioactivity [40]. Albuferon (INF α2b and HSA fusion protein, Alb-INF-α) was constructed by fusing INF α2b to the C-terminus of albumin [41]. The bioactivity of albumin fusion INF α2b is inevitably hindered by bulky albumin. However, due to the long plasma half-life, albuferon could be administered by 2 to 4-week intervals [42]. Unfortunately, it was found in phase III clinical trials that high dose of albuferon was associated with a higher rate of serious pulmonary adverse events [43]. Eventually, the development of Alb-INF-α was terminated. Nevertheless, this pioneer study of albumin fusion protein eased the immunogenicity concern on this new type of biological drug. Albuferon is a cytokine homologue with immunostimulatory function. Clinical trials have shown that this fusion protein drug was well tolerated following repeated doses and under high immune sensitivity conditions [43].

3.3. Granulocyte Colony Stimulating Factor (G-CSF)

G-CSF is used to treat neutropenia for cancer patients receiving intensive chemotherapy. G-CSF is a lineage-specific cytokine that can induce proliferation of neutrophil colonies and differentiation of neutrophil precursor cells, and promote the activity of mature neutrophils [44]. Recombinant G-CSF (filgrastim, rhG-CSF) is produced in *E. coli* and has a short plasma half-life. Therefore, daily subcutaneous injections throughout the chemotherapy cycle are needed. Neulasta (pegfilgrastim, PEGylated filgrastim) is a PEGlyated long-lasting analogue of filgrastim by attaching 20 kDa PEG to the N-terminal methionine of G-CSF [45]. Balugrastim, an albumin and G-CSF fusion protein, was constructed by fusing G-CSF to the C-terminal albumin and expressed in yeast *Saccharomyces cerevisiae*. In a phase III clinical trial, the efficacy and safety between balugrastim and pegfilgrastim have been compared in breast cancer patients receiving chemotherapy [46]. This study revealed that balugrastim was not inferior to pegfilgrastim, in terms of efficacy, safety, and immunogenicity. However, Teva withdrew its Biologics License Application (BLA) from the FDA in 2013 for additional studies as well as marketing strategies.

3.4. Factor IX (FIX)

FIX is a relative large therapeutic protein composed of four domains (GLA, EGF, EGF, and Trypsin) with a molecular weight of 57 kDa. Usually, hemophilia B is associated with the deficiency of FIX. As a proenzyme of the coagulation system, it is cleaved by either factor

VIIa/tissue factor or factor XIa from the contact pathway to be activated (FIXa). Generally, intravenous injection of recombinant or plasma-derived FIX is required for prophylactic disease management every 2~3 days [47]. To obtain a long-lasting formulation, an albumin and FIX fusion protein was constructed by fusing FIX to the C-terminus of albumin [48]. Due to the hindrance of bulky albumin on factor IXa, a fusion protein with a non-cleavable linker showed low specific activity in clotting assay. Subsequently, an FVIIa/tissue factor or FXIa-cleavable linker was included and tested to increase the bioactivity of FIXa. As expected, the activated FIXa was able to release from the fusion protein upon protease cleavage. Overall, the molar specific activity of the fusion protein with cleavable linker showed 10~30 times higher bioactivity than the fusion protein with a non-cleavable linker. Animal studies in rats and rabbits all indicated the prolonged plasma half-life for fusion proteins with or without cleavable linkers.

In short, albumin fusion strategy is commonly applied to therapeutic proteins that target either membrane receptors or hematological factors. Deep tissue penetration required for tumor treatment is not a critical factor that affects the therapeutic efficacy for the applications introduced above. The long-lasting pharmacokinetic gain is obvious for albumin fusion proteins. Inevitably, adding a bulky albumin protein to a therapeutic protein directly will interfere with the bioactivity and negatively impact the pharmacodynamic benefits. Therefore, the structure, expression, purification, and formulation of albumin fusion proteins must be carefully designed, optimized, and evaluated.

4. Accumulation of Albumin in Tumor

The long plasma half-life of albumin and albumin fusion protein not only is due to the proper molecular weight that exceeds the renal clearance threshold but also is related to the salvaging by FcRn that prevents albumin from intracellular degradation and transports albumin into extravascular spaces *via* transcytosis [49]. The function of FcRn on IgG recycling and transportation has been well investigated and recognized [50,51]. However, the role of FcRn on albumin recycling was claimed in 2003 [52] and the detailed interactions have just been revealed [25,53,54].

FcRn is a homologue of major histocompatibility complex I and consists of a transmembrane heavy chain non-covalently associated with the soluble β2-microglobulin light chain. Albumin and IgG bind to FcRn in a pH-dependent manner but at different sites [55]. It is believed that albumin or IgG is internalized by cells through pinocytosis and binds tightly under low pH with FcRn that predominantly locates within endosomes. When the ternary complex is recycled to the cell surface and exposed to the physiological pH of blood, albumin or IgG is released from FcRn and escapes from lysosomal degradation in hematopoietic cells and vascular endothelial cells lining. Structure/function studies indicate that albumin domain III is essential for FcRn binding facilitated by residues of domain I [54]. Therefore, it will be interesting to know the effect of fusion protein on FcRn binding and protein recycling. Although *in vitro* studies have shown that the N- or C-terminal fusion of scFv or small peptide only slightly altered the binding affinity of FcRn to albumin fusion protein [29], animal and human studies are really necessary to confirm these findings. So far, in all clinical studies, albumin fusion proteins showed much shorter plasma half-life than wild type albumin.

Wild type albumins, albumin fusion proteins, or albumins as drug carriers need to be transported across vascular endothelial cells lining to reach targets in extravascular spaces. Albondin (gp60) is the key receptor located on the plasma membrane of continuous endothelium to facilitate albumin internalization and transcytosis [56–58]. Albondin is a 60 kDa glycoprotein that binds specifically to wild type albumin and is responsible for transporting approximate 50% albumins across capillary lining through a caveolin-dependent endocytosis followed by transcytosis [59,60]. In contrast, modified albumins, such as chemically labeled, damaged, and old albumins, are bound, internalized, and degraded through gp18 and gp30 binding [61]. These two proteins function like scavenger receptors and are distributed on a variety of cells, including macrophages, fibroblast, endothelia, as well as some cancer cells [62,63]. It is likely that chemically modified albumins as drug carriers can be quickly eliminated through vascular scavenger receptors. Therefore, wild type albumins or recombinant albumins mimicking the native form, rather than chemically altered albumins, should be used as drug carriers.

Albumins transported into the extravascular spaces are able to bind to proteins overexpressed by or associated with tumor and accumulate in the tumor interstitium. SPARC (secreted protein acid and rich in cysteine) is a non-collagenous matricellular protein involved in the regulation of normal and cancer cell functions, such as proliferation, adhesion, migration, apoptosis, and tumor growth [64]. It is overexpressed in many tumor cells, but its function on tumorigenesis is controversial [65]. Studies suggested that SPARC and albondin shared the similar native albumin binding domain [66]. Moreover, clinical studies also indicated the expression of SPARC was associated with the patient's response to Abraxane, a paclitaxel-loaded albumin nanoparticle [67]. So far, no research has been performed to demonstrate that SPARC facilities the uptake of albumin in tumor cells. Interestingly, more albumin-binding proteins from the plasma membrane of cancer cells, such as hnRNPs and calreticulin, have been identified [68]. However, the roles of these proteins on the uptake of albumin are unclear.

In addition to normal physiological transportation and distribution routes, albumins tend to accumulate in solid tumors due to vasculature leakage and a missing lymphatic drainage system. Solid tumors commonly possess immature, chaotic, dilated, tortuous, and leakage vasculature. Macromolecules are able to leave the bloodstream and reach the interstitium of tumor. Because of the defected lymphatic drainage system in solid tumors, size-dependent diffusion will take control of the molecule movement. Small molecules will be quickly expelled, whereas high molecular weight molecules will be retained in the tumor interstitium. Hence, albumin fusion proteins can increase the accumulation of therapeutic protein/peptide in tumors *via* the enhanced permeation and retention effect (EPR) [69].

Tumor cells adapt altered metabolic pathways to gain survival advantages over normal cells. Tumor cells utilize albumins as a nutrient and energy source due to abnormal cellular metabolism [12]. For example, albumins make up 19% of the soluble proteins of certain breast cancer cells [70]. Thorough cellular and animal studies proved the increased rate of albumin uptake in tumor cells [71,72]. Back in 1948, it was proposed that tumors acted as 'nitrogen trap' [73]. Meanwhile, albumins contribute to around 70% overall transportable nitrogen in blood. Extensive studies have just been explored to reveal the mechanisms. A recent study showed that Ras-transformed cells acquired amino acid supply *via* macropinocytosis, a process mediated by large and heterogeneous macropinosomes to internalize extracellular fluid and its contents [74]. Internalized proteins, exemplified by FITC-labeled albumins in this study, were degraded to yield glutamines for central carbon metabolism. Active tumor cell uptake of albumins implies that albumin fusion protein can be a novel approach to deliver peptide or protein to act on intracellular targets. For instance, an albumin and p53-activating peptide fusion protein has proved the feasibility of this approach [75].

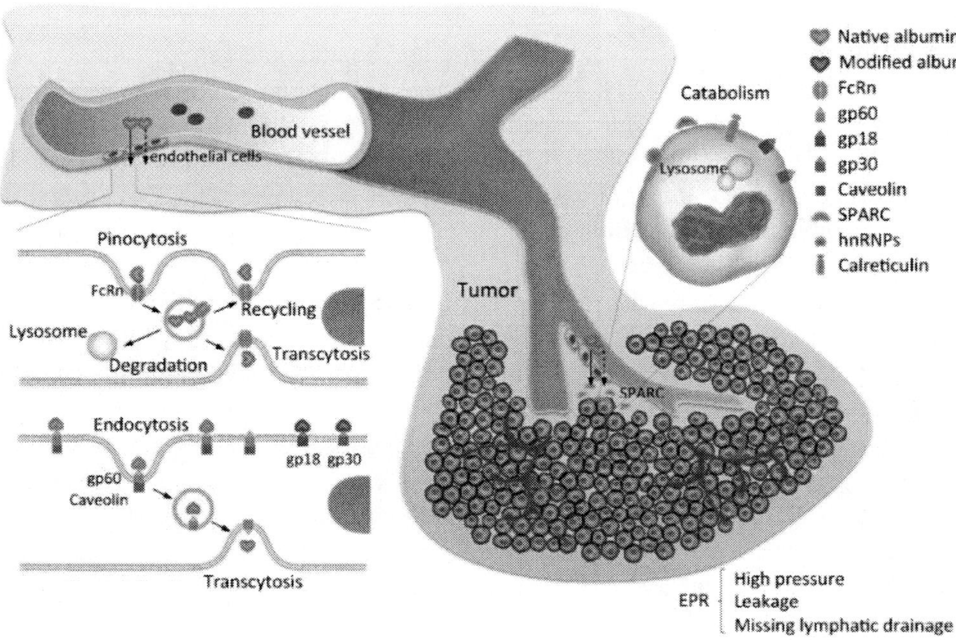

Figure 1. Schematic diagram of albumin recycling, intracellular degradation, extravascular transportation, tumor tissue accumulation, and cellular uptake. In normal tissues, albumins leave vascular endothelial cells and reach extravascular spaces mainly through FcRn-mediated pinocytosis and transcytosis or caveolin-dependent gp60 (albondin)-mediated endocytosis and transcytosis. Binding to FcRn allows albumin to escape from endosome-lysosome degradation, thus contributes to the long plasma half-life. Unlike gp60 that shows high affinity to native albumin, gp18 and gp30 bind preferably to modified albumin and result in protein internalization and degradation. In tumor tissues, albumin and other macromolecules tend to accumulate in the tumor interstitium due to EPR (enhanced permeation and retention effect). Several albumin binding proteins associated with various types of cancer cells, such as SPARC, hnRNPs, and calreticulin, have been identified with unclear functions. Finally, albumins are actively taken by tumor cells as a nutrient source.

p53 protein is an essential tumor suppressor involved in intracellular stress signals, including DNA damage, oncogene activation, and hypoxia, and regulates cellular apoptosis, proliferation and senesce [76]. Degradation of p53 is regulated by MDM2, an ubiquitin E3 ligase. Disrupting the interaction between p53 and MDM2 has been a well-studied approach to restore p53 population and induce apoptosis in tumor cells [77]. Nutlin and other small molecules only inhibit p53 and MDM2 interaction [78]. MDMX, a MDM2 homologue, can compensate the MDM2 inhibition. Peptide-based approaches can inhibit both MDM2 as well as MDMX [79]; however, peptide delivery and cell penetration are problematic.

The p53-activating peptide derived from the N-terminus of p53 (residues 17~28) can block the interaction between p53 and MDM2/MDMX. This p53-activating peptide was fused to the C-terminus of albumin (rHSA-p53i) connected by a caspase 3 cleavable linker (DEVDG) [75]. This albumin fusion protein could be efficiently taken up by SJSA-1 cells *in vitro*, interacted with MDM2 as well as MDMX, and resulted in p53 accumulation. Subsequently, apoptosis and cell death have been induced. An ongoing research using SJSA-1 tumor cell xenograft mouse model has shown that combination treatment of methotrexate with rHSA-p53i could significantly halt tumor growth compared with methotrexate sole treatment. It seems that the utilization of albumin fusion protein is a feasible and effective approach to target an essential intracellular pathway for cancer therapy. Figure 1 presents the processes involved in albumin recycling, degradation, extravascular transportation, tumor tissue accumulation, and cellular uptake. All these processes may affect the pharmacokinetic profile and pharmacodynamic benefit of albumin-based drugs.

5. Albumin Fusion Proteins As Drug Carriers

In addition to fusion protein strategy, covalently conjugating of chemicals to albumin is also a common approach to develop a long-lasting formulation. The anti-tumor and anti-arthritis efficacy of methotrexate-conjugated HSA (MTX-HSA), for instance, has been evaluated in phase I and II clinical trials [80]. Many other types of drug-conjugated HSA have also been tested. Originally, it was thought that one albumin could carry multiple covalently conjugated payloads to enhance cytotoxicity. This approach might show optimal *in vitro* cytotoxicity; however, considering the overall *in vivo* pharmacokinetic profile, the 1.3 payload to albumin ratio resulted in an optimal therapeutic efficacy [81]. Because gp18 and gp30 preferably bind and mediate the scavenging of conformationally modified albumin, albumin with 1 or 2 conjugations may thereby mimic the native albumin and is able to escape from degradation. Obviously, when utilizing albumins as drug carriers, it is important to preserve the native structure and function of albumin. Alternatively, a new type of prodrug has been developed to conjugate endogenous albumin specifically at the free Cys34 residue. The payload to albumin ratio is 1 due to the sole modification site. DOXO-EMCH binds to the free thiol group of albumin *in vivo*, through a pH sensitive bond [82]. Upon reaching the site of target and being internalized by endocytosis, the drug can be released when the pH changes. Since the free thiol group of albumin is critical for its biological function, the effect of Cys34 modification on the fate of albumin circulation should be elucidated.

Formulating albumins and drugs into nanoparticles or microspheres has been well-developed. Abraxane is an albumin nanoparticle encapsulating paclitaxel produced by the nab-technology under high air pressure [83]. This technology takes advantage of the high lipophilicity of paclitaxel and its tight binding with albumin. Nanoparticle formulation could also be applied to albumin fusion proteins with target-binding function for targeted drug delivery. However, retaining the active structure of albumin fusion protein during manufacturing could be problematic. Moreover, chemical conjugation or sophisticated nano particle manufacturing may cause the instability of formulation, increase the rate of plasma clearance mediated by scavenger receptors, and boost the chance of immunoreactions during treatments. Compelxing albumin or albumin fusion protein with payloads by non-covalent binding may minimize the alteration of albumin structure and function and benefit drug distribution, retention, and uptake.

A variety of methods, such as phage display, affibody, and antibody, have been applied to screen polypeptides that bind tightly to albumin. Then, payloads can be conjugated to albumin binding peptides or therapeutic protein/peptide can be fused to albumin binding peptides. It is expected that albumin binding peptides and its payloads form tight complexes with endogenous albumins and subsequently gain a preferred pharmacological profile. AlbudAbs is a HSA binding antibody developed by GSK [84], whereas Albumod is an affibody molecule with high albumin binding affinity [85]. Interestingly, it seems feasible to form albumin fusion protein and albumin binding peptide/payload complex *in vitro* to obtain multifunctional albumin complexes for targeted therapy or synergetic combination therapy. Nevertheless, synthesis of albumin binding peptide and payload conjugates may face many challenges during manufacturing.

One main biological function of albumin is to transport LCFAs. Compelxing albumin or albumin fusion protein with fatty acid-modified drugs *in vitro* followed by *in vivo* administration is an easy and non-covalent approach that takes advantages of the physiological functions of albumin. LCFA is an essential nutrient and energy source for many cell types. One albumin molecule possesses 8 putative LCFAs or MCFAs (medium chain fatty acid) binding sites and is capable of binding up to 6 LCFAs simultaneously with 10^{-8}~10^{-6} M binding affinity [86]. LCFA-modified drugs are able to form stable complexes with albumins and albumin fusion proteins. Therefore, LCFA-modified drugs can circulate longer in plasma and may be preferably taken by tumor cells as a nutrient and energy source.

So far, few fatty acid-modified protein drugs have been successfully developed. Insulin detemir [87] and liraglutide [88] are myristic acid (C_{14})-modified insulin and GLP-1, respectively. Upon injection, fatty acid-modified drugs bind to endogenous albumin circulating in the bloodstream and gain significant extension of serum half-life. Many fatty acid-modified small molecule drugs are under development. A recent study has reported that linear alkyl groups could be used to modify a Pt(IV) cisplatin-derived prodrug and then form non-covalent drug/albumin complexes as discussed above [89]. These complexes showed significant improvement in cytotoxicity compared with cisplatin as well as an extended half-life in serum.

Instead of forming an albumin and drug complex randomly *in vivo*, albumin and fatty acid-modified drugs can form complexes under optimal conditions *in vitro*. It allows the opportunity to optimize the payload to albumin ratio and make complexes that carry different types of drugs for combination therapy. More importantly, albumin fusion proteins can also be formulated to carry fatty acid-modified drugs.

Albumin and p53-activating peptide fusion protein disrupted the interaction between p53 and MDM2 and induced apoptosis [75]. Unlike chemically modified albumin, this fusion protein maintained the proper protein folding as wild type albumin manifested by comparable fatty acid binding capability. Similar to the wild type albumin, one rHSA-p53i fusion protein was able to bind to up to 4 fatty acid-modified FITC (FA-FITC) *in vitro*. Albumin-bound FA-FITC was really stable and could not be replaced by 10% serum after 24 hours of incubation. Albumin and drug complexes were efficiently taken by SJSA-1 cells *in vitro*.

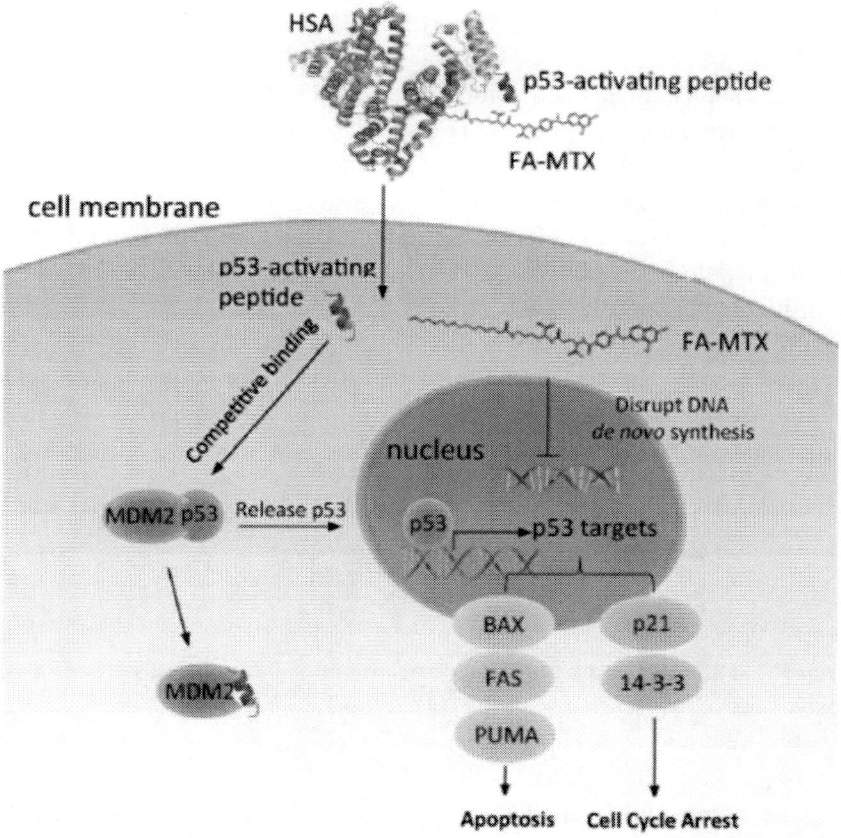

Figure 2. Schematic diagram of albumin fusion protein-mediated co-delivery technology. A p53-activating peptide is fused to the C-terminus of HSA (rHSA-p53i). The albumin fusion protein is then complexed with fatty acid modified methotrexate (FA-MTX) *in vitro*. The drug complexes transport both p53–activating peptide and MTX into the same tumor cell via multiple mechanisms. Inside the tumor cell, MTX causes DNA damage and subsequently induces apoptosis. Meanwhile, p53-activating peptide disrupts the interaction between p53 and MDM2/MDMX and boosts the cellular apoptotic response to DNA damage. Synergistic cytotoxicity is achieved by co-delivering of two drugs into one tumor cell.

Because rHSA-p53i is capable of restoring the normal p53 cellular function and sensitizing tumor cells to cytotoxicities caused by chemotherapeutic drugs, co-delivery of one or more drugs into one tumor cell *in vivo* with rHSA-p53i can achieve synergistic antitumor efficacy and minimize systemic toxicity. Albumin fusion protein, rHSA-p53i, complexed with fatty acid-modified chemotherapeutics, assures two drugs to act on two different cellular pathways in one cell simultaneously, as shown in Figure 2. rHSA-p53i complexed with fatty acid-modified methotrexate (FA-MTX) did show synergistic cytotoxicity in SJSA-1 tumor cell xenograft mouse. Furthermore, this strategy not only improves the therapeutic efficacy but also limits opportunities to develop chemoresistance. It fully takes advantage of the physiological function of wild type albumin to minimize drug clearance mediated by scavenger receptors and maximize drug uptake in tumor cells. Lastly, this *in vitro* non-covalent FA-based formulation approach is easy to manufacture and scale-up.

Conclusion

Despite tremendous studies and efforts over the last few decades, few albumin-based drugs have been successfully developed. Taking advantage of cutting-edge biotechnologies, many unsolved or uncertain questions on albumin, such as FcRn binding and macropinocytosis, have been revisited and revealed. Fully understanding the biological functions of albumin and mechanisms supporting these functions will definitely facilitate the developing of albumin-based drugs. Of course, these drugs should mimic the wild type albumin to fulfill the pharmacokinetic benefits.

Albumin fusion protein is not only a feasible strategy to improve the plasma half-life of therapeutic proteins and peptides but also an efficient carrier for intracellular peptide delivery and targeting. Taking advantage of the high affinity and multisite FA binding and transportation, albumin fusion protein can be utilized to deliver FA-modified drugs for synergistic cytotoxicity in cancer therapy.

References

[1] Caironi P, Gattinoni L. The clinical use of albumin: the point of view of a specialist in intensive care. *Blood Transfus*. 2009;7(4):259–67.

[2] Kratz F. Albumin as a drug carrier: design of prodrugs, drug conjugates and nanoparticles. *J. Control. Release*. 2008;132(3):171–83.

[3] Sleep D. Albumin and its application in drug delivery. *Expert Opin. Drug Deliv.* 2015 Dec 18; 12(5): 793-812.

[4] Rogers B, Dong D, Li Z, Li Z. Recombinant human serum albumin fusion proteins and novel applications in drug delivery and therapy. *Curr. Pharm. Des*. 2015; 21(14):1899–907.

[5] Schulte S. Half-life extension through albumin fusion technologies. *Thromb. Res*. 2009;124 Suppl S6–8.

[6] Shimanovich U, Bernardes GJ, Knowles TP–, Cavaco-Paulo A. Protein micro- and nano-capsules for biomedical applications. *Chem. Soc. Rev.* 2014;43(5):1361–71.

[7] Fishburn CS. The pharmacology of PEGylation: balancing PD with PK to generate novel therapeutics. *J. Pharm. Sci*. 2008;97(10):4167–83.

[8] Elliott S, Lorenzini T, Asher S, Aoki K, Brankow D, Buck L, et al. Enhancement of therapeutic protein in vivo activities through glycoengineering. *Nat. Biotechnol.* 2003;21(4):414–21.

[9] Schmidt SR. Fusion-proteins as biopharmaceuticals--applications and challenges. *Curr. Opin. Drug Discov. Devel.* 2009;12(2):284–95.

[10] Fischer A. Press Announcements - FDA approves Tanzeum to treat type 2 diabetes. Center for Drug Evaluation and Research; [cited 2015 Apr 11].http://www.fda.gov/NewsEvents/Newsroom/PressAnnouncements/ucm393289.htm

[11] Quinlan GJ, Martin GS, Evans TW. Albumin: biochemical properties and therapeutic potential. *Hepatology*. 2005;41(6):1211–9.

[12] Stehle G, Sinn H, Wunder A, Schrenk HH, Stewart JCM, Hartung G, et al. Plasma protein (albumin) catabolism by the tumor itself—implications for tumor metabolism and the genesis of cachexia. Crit. *Rev. Oncol. Hematol.* 1997;26(2):77–100.

[13] Shet AS, Jahagirdar BN, Verfaillie CM. Chronic myelogenous leukemia: mechanisms underlying disease progression. 2002;16(8):1402–11.

[14] Czajkowsky DM, Hu J, Shao Z, Pleass RJ. Fc-fusion proteins: new developments and future perspectives. *EMBO Mol. Med.* 2012; 4(10):1015–28.

[15] Brandsma ME, Jevnikar AM, Ma S. Recombinant human transferrin: beyond iron binding and transport. *Biotechnol. Adv.* 2011;29(2):230–8.

[16] Sleep D, Cameron J, Evans LR. Albumin as a versatile platform for drug half-life extension. *Biochim. Biophys. Acta*. 2013;1830(12):5526–34.

[17] De Bruyn M, Bremer E, Helfrich W. Antibody-based fusion proteins to target death receptors in cancer. *Cancer Lett.* 2013;332(2):175–83.

[18] Bagshawe KD, Sharma SK, Begent RH. Antibody-directed enzyme prodrug therapy (ADEPT) for cancer. *Expert Opin. Biol. Ther.* 2004;4(11):1777–89.

[19] Ortiz-Sánchez E, Helguera G, Daniels TR, Penichet ML. Antibody-cytokine fusion proteins: applications in cancer therapy. *Expert Opin. Biol. Ther.* 2008;8(5):609–32.

[20] LeBowitz JH, Grubb JH, Maga JA, Schmiel DH, Vogler C, Sly WS. Glycosylation-independent targeting enhances enzyme delivery to lysosomes and decreases storage in mucopolysaccharidosis type VII mice. *Proc. Natl. Acad. Sci. U. S. A.* 2004;101(9):3083–8.

[21] Medina-Echeverz J, Fioravanti J, Díaz-Valdés N, Frank K, Aranda F, Gomar C, et al. Harnessing high density lipoproteins to block transforming growth factor beta and to inhibit the growth of liver tumor metastases. *PLoS One*. 2014;9(5):e96799.

[22] Duvic M, Talpur R. Optimizing denileukin diftitox (Ontak®) therapy. *Future Oncol.* 2008;4(4):457–69.

[23] Sockolosky JT, Szoka FC. The neonatal Fc receptor, FcRn, as a target for drug delivery and therapy. *Adv. Drug Deliv. Rev.* 2015; doi:10.1016/j.addr.2015.02.005

[24] Desjarlais JR, Lazar GA. Modulation of antibody effector function. *Exp. Cell Res.* 2011;317(9):1278–85.

[25] Sand KM, Bern M, Nilsen J, Noordzij HT, Sandlie I, Andersen JT. Unraveling the interaction between FcRn and albumin: Opportunities for design of albumin-based therapeutics. *Front Immunol.* 2015; 682.

[26] Kawakami A, Kubota K, Yamada N, Tagami U, Takehana K, Sonaka I, et al. Identification and characterization of oxidized human serum albumin. A slight structural change impairs its ligand-binding and antioxidant functions. *FEBS J.* 2006; 273(14):3346–57.

[27] Peters T. All About Albumin:Biochemistry, Genetics, and Medical Applications. 1st ed. Academic Press; 1995.

[28] Anderson CL, Chaudhury C, Kim J, Bronson CL, Wani MA, Mohanty S. Perspective-- FcRn transports albumin: relevance to immunology and medicine. *Trends Immunol.* 2006; 27(7):343–8.

[29] Andersen JT, Cameron J, Plumridge A, Evans L, Sleep D, Sandlie I. Single-chain variable fragment albumin fusions bind the neonatal Fc receptor (FcRn) in a species-dependent manner: implications for *in vivo* half-life evaluation of albumin fusion therapeutics. *J. Biol. Chem.* 2013; 288(33):24277–85.

[30] Zhao J, Si Y, Cheng M, Yang Y, Niu Y, Li X, et al. Albumin fusion of interleukin-28B: production and characterization of its biological activities and protein stability. *PLoS One*. 2013; 8(5):e64301.

[31] Zhao HL, Xue C, Wang Y, Li XY, Xiong XH, Yao XQ, et al. Circumventing the heterogeneity and instability of human serum albumin-interferon-alpha2b fusion protein by altering its orientation. *J. Biotechnol.* 2007; 131(3):245–52.

[32] Zhao HL, Yao XQ, Xue C, Wang Y, Xiong XH, Liu ZM. Increasing the homogeneity, stability and activity of human serum albumin and interferon-alpha2b fusion protein by linker engineering. *Protein Expr. Purif.* 2008; 61(1):73–7.

[33] Chen X, Zaro JL, Shen WC. Fusion protein linkers: property, design and functionality. *Adv. Drug Deliv. Rev.* 2013; 65(10):1357–69.

[34] Doyle ME, Egan JM. Mechanisms of action of glucagon-like peptide 1 in the pancreas. *Pharmacol. Ther.* 2007; 113(3):546–93.

[35] Barnett A. Exenatide. *Expert Opin. Pharmacother.* 2007; 8(15):2593–608.

[36] Malone J, Trautmann M, Wilhelm K, Taylor K, Kendall DM. Exenatide once weekly for the treatment of type 2 diabetes. *Expert Opin. Investig. Drugs*. 2009; 18(3):359–67.

[37] Matthews JE, Stewart MW, De Boever EH, Dobbins RL, Hodge RJ, Walker SE, et al. Pharmacodynamics, pharmacokinetics, safety, and tolerability of albiglutide, a long-acting glucagon-like peptide-1 mimetic, in patients with type 2 diabetes. *J. Clin. Endocrinol. Metab.* 2008; 93(12):4810–7.

[38] Jaitin DA, Roisman LC, Jaks E, Gavutis M, Piehler J, Van der Heyden J, et al. Inquiring into the differential action of interferons (IFNs): an IFN-alpha2 mutant with enhanced affinity to IFNAR1 is functionally similar to IFN-beta. *Mol. Cell. Biol.* 2006; 26(5):1888–97.

[39] Foser S, Schacher A, Weyer KA, Brugger D, Dietel E, Marti S, et al. Isolation, structural characterization, and antiviral activity of positional isomers of monopegylated interferon α-2a (PEGASYS). *Protein Expr. Purif.* 2003; 30(1):78–87.

[40] Wang YS, Youngster S, Grace M, Bausch J, Bordens R, Wyss DF. Structural and biological characterization of pegylated recombinant interferon alpha-2b and its therapeutic implications. *Adv. Drug Deliv. Rev.* 2002; 54(4):547–70.

[41] Osborn BL, Olsen HS, Nardelli B, Murray JH, Zhou JOEXH, Garcia A, et al. Pharmacokinetic and pharmacodynamic studies of a human serum albumin-interferon-alpha fusion protein in cynomolgus monkeys. *J. Pharmacol. Exp. Ther.* 2002;303(2):540–8.

[42] Subramanian GM, Fiscella M, Lamousé-Smith A, Zeuzem S, McHutchison JG. Albinterferon alpha-2b: a genetic fusion protein for the treatment of chronic hepatitis C. *Nat. Biotechnol.* 2007; 25(12):1411–9.

[43] Zeuzem S, Sulkowski MS, Lawitz EJ, Rustgi VK, Rodriguez-Torres M, Bacon BR, et al. Albinterferon Alfa-2b was not inferior to pegylated interferon-α in a randomized trial of patients with chronic hepatitis C virus genotype 1. *Gastroenterology*. 2010;139(4):1257–66.

[44] Root RK, Dale DC. Granulocyte colony-stimulating factor and granulocyte-macrophage colony-stimulating factor: comparisons and potential for use in the treatment of infections in nonneutropenic patients. *J. Infect. Dis.* 1999;179 Suppl 2:S342–52.

[45] Morishita M, Leonard RC. Pegfilgrastim; a neutrophil mediated granulocyte colony stimulating factor-expanding uses in cancer chemotherapy. *Expert Opin. Biol. Ther.* 2008;8(7):993–1001.

[46] Volovat C, Gladkov OA, Bondarenko IM, Barash S, Buchner A, Bias P, et al. Efficacy and safety of balugrastim compared with pegfilgrastim in patients with breast cancer receiving chemotherapy. *Clin. Breast Cancer.* 2014;14(2):101–8.

[47] Björkman S, Shapiro AD, Berntorp E. Pharmacokinetics of recombinant factor IX in relation to age of the patient: implications for dosing in prophylaxis. *Haemophilia.* 2001;7(2):133–9.

[48] Metzner HJ, Weimer T, Kronthaler U, Lang W, Schulte S. Genetic fusion to albumin improves the pharmacokinetic properties of factor IX. *Thromb. Haemost.* 2009;102(4):634–44.

[49] Andersen JT, Dalhus B, Cameron J, Daba MB, Plumridge A, Evans L, et al. Structure-based mutagenesis reveals the albumin-binding site of the neonatal Fc receptor. *Nat. Commun.* 2012;3:610.

[50] Lencer WI, Blumberg RS. A passionate kiss, then run: exocytosis and recycling of IgG by FcRn. *Trends Cell Biol.* 2005;15(1):5–9.

[51] Hinton PR, Xiong JM, Johlfs MG, Tang MT, Keller S, Tsurushita N. An engineered human IgG1 antibody with longer serum half-life. *J. Immunol.* 2006;176(1):346–56.

[52] Chaudhury C, Mehnaz S, Robinson JM, Hayton WL, Pearl DK, Roopenian DC, et al. The major histocompatibility complex-related Fc receptor for IgG (FcRn) binds albumin and prolongs its lifespan. J. *Exp. Med.* 2003;197(3):315–22.

[53] Schmidt MM, Townson SA, Andreucci AJ, King BM, Schirmer EB, Murillo AJ, et al. Crystal structure of an HSA/FcRn somplex reveals recycling by competitive mimicry of HSA ligands at a pH-dependent hydrophobic interface. *Structure.* 2013;21(11):1966–78.

[54] Sand KMK, Bern M, Nilsen J, Dalhus B, Gunnarsen KS, Cameron J, et al. Interaction with both domain I and III of albumin is required for optimal pH-dependent binding to the neonatal Fc receptor (FcRn). *J. Biol. Chem.* 2014;289(50):34583–94.

[55] Chaudhury C, Brooks CL, Carter DC, Robinson JM, Anderson CL. Albumin binding to FcRn: distinct from the FcRn-IgG interaction. *Biochemistry.* 2006;45(15):4983–90.

[56] Schnitzer JE, Oh P. Albondin-mediated capillary permeability to albumin. Differential role of receptors in endothelial transcytosis and endocytosis of native and modified albumins. *J. Biol. Chem.* 1994;269(8):6072–82.

[57] Tiruppathi C, Song W, Bergenfeldt M, Sass P, Malik AB. Gp60 activation mediates albumin transcytosis in endothelial cells by tyrosine kinase-dependent pathway. *J. Biol. Chem.* 1997;272(41):25968–75.

[58] Vogel SM, Minshall RD, Pilipovic M, Tiruppathi C, Malik AB, Stephen M, et al. Albumin uptake and transcytosis in endothelial cells *in vivo* induced by albumin-binding protein. *Am. J. Physiol. Lung Cell Mol Physiol.* 2001;281:1512–22.

[59] John TA, Vogel SM, Minshall RD, Ridge K, Tiruppathi C, Malik AB. Evidence for the role of alveolar epithelial gp60 in active transalveolar albumin transport in the rat lung. *J. Physiol.* 2001; 533(Pt 2):547–59.

[60] Merlot AM, Kalinowski DS, Richardson DR. Unraveling the mysteries of serum albumin-more than just a serum protein. *Front Physiol.* 2014;5:299.

[61] Schnitzer JE, Bravo J. High affinity binding, endocytosis, and degradation of conformationally modified albumins. Potential role of gp30 and gp18 as novel scavenger receptors. *J. Biol. Chem.* 1993;268(10):7562–70.

[62] Schnitzer JE. Gp60 is an albumin-binding glycoprotein expressed by continuous endothelium involved in albumin transcytosis. *Am. J. Physiol. Hear. Circ. Physiol.* 1992;262(1):H246–254.

[63] Wang J, Ueno H, Masuko T, Hashimoto Y. Binding of serum albumin on tumor cells and characterization of the albumin binding protein. *J. Biochem.* 1994;115(5):898–903.

[64] Nagaraju GP, Dontula R, El-Rayes BF, Lakka SS. Molecular mechanisms underlying the divergent roles of SPARC in human carcinogenesis. *Carcinogenesis.* 2014;35(5):967–73.

[65] Chen J, Wang M, Xi B, Xue J, He D, Zhang J, et al. SPARC is a key regulator of proliferation, apoptosis and invasion in human ovarian cancer. *PLoS One.* 2012;7(8):e42413.

[66] Schnitzer JE, Oh P. Antibodies to SPARC inhibit albumin binding to SPARC, gp60, and microvascular endothelium. *Am J Physiol Hear. Circ Physiol.* 1992;263(6):H1872–1879.

[67] Desai N, Trieu V, Damascelli B, Soon-Shiong P. SPARC expression correlates with tumor response to albumin-bound paclitaxel in head and neck cancer patients. *Transl. Oncol.* 2009;2(2):59–64.

[68] Fritzsche T, Schnölzer M, Fiedler S, Weigand M, Wiessler M, Frei E. Isolation and identification of heterogeneous nuclear ribonucleoproteins (hnRNP) from purified plasma membranes of human tumour cell lines as albumin-binding proteins. *Biochem. Pharmacol.* 2004;67(4):655–65.

[69] Maeda H, Wu J, Sawa T, Matsumura Y, Hori K. Tumor vascular permeability and the EPR effect in macromolecular therapeutics: a review. *J. Control. Release.* 2000;65(1-2):271–84.

[70] Søreide JA, Lea OA, Kvinnsland S. Cytosol albumin content in operable breast cancer. Correlations to steroid hormone receptors, other prognostic factors and prognosis. *Acta Oncol.* 1991;30(7):797–802.

[71] Sinn H, Schrenk HH, Friedrich EA, Schilling U, Maier-Borst W. Design of compounds having an enhanced tumour uptake, using serum albumin as a carrier. *Part I. Int. J. Rad. Appl. Instrum.* B. 1990;17(8):819–27.

[72] Schilling U, Friedrich EA, Sinn H, Schrenk HH, Clorius JH. Design of Compounds Having Enhanced Tumour Uptake , Using Serum Albumin as a Carrier-Part II . In Vivo Studies. *Int. J. Rad. Appl. Instrum B.* 1992;19(6):685–95.

[73] Wunder A, Stehle G, Sinn H, Schrenk H, Hoffbiederbeck D, Bader F, et al. Enhanced albumin uptake by rat tumors. *Int. J. Oncol.* 1997;11:497–507.

[74] Commisso C, Davidson SM, Soydaner-Azeloglu RG, Parker SJ, Kamphorst JJ, Hackett S, et al. Macropinocytosis of protein is an amino acid supply route in Ras-transformed cells. *Nature.* 2013;497 (7451):633–7.

[75] Joshi MR, Yao N, Myers KA, Li Z. Human serum albumin and p53-activating peptide fusion protein is able to promote apoptosis and deliver fatty acid-modified molecules. *PLoS One.* 2013;8(11):e80926.

[76] Prives C, Hall PA. The p53 pathway. *J. Pathol.* 1999;187(1):112–26.

[77] Chen F, Wang W, El-Deiry WS. Current strategies to target p53 in cancer. *Biochem. Pharmacol.* 2010;80(5):724–30.

[78] Fischer PM, Lane DP. Small-molecule inhibitors of the p53 suppressor HDM2: have protein-protein interactions come of age as drug targets? *Trends Pharmacol. Sci.* 2004;25(7):343–6.

[79] Kanovsky M, Raffo A, Drew L, Rosal R, Do T, Friedman FK, et al. Peptides from the amino terminal mdm-2-binding domain of p53, designed from conformational analysis, are selectively cytotoxic to transformed cells. *Proc. Natl. Acad. Sci. U. S. A.* 2001;98(22):12438–43.

[80] Bolling C, Graefe T, Lübbing C, Jankevicius F, Uktveris S, Cesas A, et al. Phase II study of MTX-HSA in combination with cisplatin as first line treatment in patients with advanced or metastatic transitional cell carcinoma. *Invest. New Drugs.* 2006;24(6):521–7.

[81] Stehle G, Wunder A, Sinn H, Schrenk HH, Schütt S, Frei E, et al. Pharmacokinetics of methotrexate-albumin conjugates in tumor-bearing rats. *Anticancer Drugs.* 1997;8(9):835–44.

[82] Kratz F, Azab S, Zeisig R, Fichtner I, Warnecke A. Evaluation of combination therapy schedules of doxorubicin and an acid-sensitive albumin-binding prodrug of doxorubicin in the MIA PaCa-2 pancreatic xenograft model. *Int. J. Pharm.* 2013;441(1-2):499–506.

[83] Hoy SM. Albumin-bound Paclitaxel: a review of its use for the first-line combination treatment of metastatic pancreatic cancer. *Drugs.* 2014;74(15):1757–68.

[84] Holt LJ, Basran A, Jones K, Chorlton J, Jespers LS, Brewis ND, et al. Anti-serum albumin domain antibodies for extending the half-lives of short lived drugs. *Protein Eng. Des. Sel.* 2008;21(5):283–8.

[85] Nilvebrant J, Alm T, Hober S, Löfblom J. Engineering bispecificity into a single albumin-binding domain. *PLoS One.* 2011;6(10):e25791.

[86] Curry S, Brick P, Franks NP. Fatty acid binding to human serum albumin: new insights from crystallographic studies. *Biochim. Biophys. Acta.* 1999;1441(2-3):131–40.

[87] Home P, Kurtzhals P. Insulin detemir: from concept to clinical experience. *Expert Opin. Pharmacother.* 2006 Feb 31;7(3):325–43.

[88] Ladenheim E. Liraglutide and obesity: a review of the data so far. *Drug Des. Devel. Ther.* 2015;9:1867.

[89] Zheng Y, Suntharalingam K, Johnstone TC, Yoo H, Lin W, Brooks JG, et al. Pt(IV) prodrugs designed to bind non-covalently to human serum albumin for drug delivery. *J. Am. Chem. Soc.* 2014;136(24):8790–8.

[90] Kim Y, Lee SM, Chung H. Novel AGLP-1 albumin fusion protein as a long-lasting agent for type 2 diabetes. *BMB Rep.* 2013;46(12):606–10.

[91] Huang Y, Chen ZHI, Chen Y, Ma G, Shan J. Preparation and characterization of a novel exendin-4 human serum albumin fusion protein expressed in Pichia pastoris. *J. Pept. Sci.* 2008;14(5):588–95.

[92] Halpern W, Riccobene TA, Agostini H, Baker K, Stolow D, Gu M, et al. Albugranin, a recombinant human granulocyte colony stimulating factor (g-csf) genetically fused to recombinant human albumin induces prolonged myelopoietic effects in mice and monkeys. *Pharm. Res.* 2002;19(11):1720–9.

[93] Lei J, Guan B, Li B, Duan Z, Chen Y, Li H, et al. Expression, purification and characterization of recombinant human interleukin-2-serum albumin (rhIL-2-HSA) fusion protein in Pichia pastoris. *Protein Expr. Purif.* 2012;84(1):154–60.
[94] Duttaroy A, Kanakaraj P, Osborn BL, Schneider H, Pickeral OK, Chen C, et al. Development of a long-acting insulin analog using albumin fusion technology. *Diabetes.* 2005;54:251–8.
[95] Osborn BL, Sekut L, Corcoran M, Poortman C, Sturm B, Chen G, et al. Albutropin: a growth hormone-albumin fusion with improved pharmacokinetics and pharmacodynamics in rats and monkeys. *Eur. J. Pharmacol.* 2002;456(1-3):149–58.
[96] Scehnet JS, Ley EJ, Krasnoperov V, Liu R, Manchanda PK, Sjoberg E, et al. The role of Ephs , Ephrins , and growth factors in Kaposi sarcoma and implications of EphrinB2 blockade. *Blood.* 2009;113(1):254–63.
[97] Huang YJ, Lundy PM, Lazaris A, Huang Y, Baldassarre H, Wang B, et al. Substantially improved pharmacokinetics of recombinant human butyrylcholinesterase by fusion to human serum albumin. *BMC Biotechnol.* 2008;8:50.
[98] Wang W, Ou Y, Shi Y. AlbuBNP, a recombinant b-type natriuretic peptide and human serum albumin fusion hormone, as a long-term therapy of congestive heart failure. *Pharm. Res.* 2004;21(11):2105–11.
[99] Joung C-H, Shin J-Y, Koo J-K, Lim JJ, Wang J-S, Lee S-J, et al. Production and characterization of long-acting recombinant human albumin-EPO fusion protein expressed in CHO cell. *Protein Expr. Purif.* 2009;68(2):137–45.
[100] Marques JA, George JK, Smith IJ, Bhakta V, Sheffield WP. A barbourin-albumin fusion protein that is slowly cleared *in vivo* retains the ability to inhibit platelet aggregation *in vitro*. *Thromb Haemost.* 2001;86(3):902–8.
[101] Sheffield WP, Smith IJ, Syed S, Bhakta V. Prolonged *in vivo* anticoagulant activity of a hirudin-albumin fusion protein secreted from Pichia pastoris. *Blood Coagul. Fibrinolysis.* 2001;12(6):433–43.
[102] Weimer T, Wormsbächer W, Kronthaler U, Lang W, Liebing U, Schulte S. Prolonged in-vivo half-life of factor VIIa by fusion to albumin. *Thromb. Haemost.* 2008;99(4):659–67.
[103] Hagedorn I, Schmidbauer S, Pleines I, Kleinschnitz C, Kronthaler U, Stoll G, et al. Factor XIIa inhibitor recombinant human albumin Infestin-4 abolishes occlusive arterial thrombus formation without affecting bleeding. *Circulation.* 2010;121(13):1510–7.
[104] Ikuta S, Chuang VT, Ishima Y, Nakajou K, Furukawa M, Watanabe H, et al. Albumin fusion of thioredoxin--the production and evaluation of its biological activity for potential therapeutic applications. *J. Control. Release.* 2010;147(1):17–23.
[105] Chen J-H, Zhang X-G, Jiang Y-T, Yan L-Y, Tang L, Yin Y-W, et al. Bioactivity and pharmacokinetics of two human serum albumin-thymosin alpha1-fusion proteins, rHSA-Talpha1 and rHSA-L-Talpha1, expressed in recombinant Pichia pastoris. *Cancer Immunol. Immunother.* 2010;59(9):1335–45.
[106] Müller N, Schneider B, Pfizenmaier K, Wajant H. Superior serum half life of albumin tagged TNF ligands. *Biochem. Biophys. Res. Commun.* 2010;396(4):793–9.
[107] Zhan J, Chen Y, Yuan H-Y, Li H, Lu H. Fusion of HSA influences TNF-α neutralizing activity of shTNFRs. *Biotechnol. Lett.* 2012;34(3):417–23.

[108] Evans L, Hughes M, Waters J, Cameron J, Dodsworth N, Tooth D, et al. The production, characterisation and enhanced pharmacokinetics of scFv-albumin fusions expressed in Saccharomyces cerevisiae. *Protein Expr. Purif.* 2010;73(2):113–24.
[109] Müller D, Karle A, Meissburger B, Höfig I, Stork R, Kontermann RE. Improved pharmacokinetics of recombinant bispecific antibody molecules by fusion to human serum albumin. *J. Biol. Chem.* 2007;282(17):12650–60.
[110] Yao Z, Dai W, Perry J, Brechbiel MW, Sung C. Effect of albumin fusion on the biodistribution of interleukin-2. *Cancer Immunol. Immunother.* 2004;53(5):404–10.

In: Human Serum Albumin (HSA)
Editor: Travis Stokes
ISBN: 978-1-63482-963-2

Chapter 4

PHARMACOKINETICALLY MODIFIED HUMAN SERUM ALBUMIN BASED THERAPEUTIC DESIGN AND DEVELOPMENT

***Kazuaki Taguchi*[1], *Victor Tuan Giam Chuang*[2], *Keishi Yamasaki*[1,3] *and Masaki Otagiri*[1,3,*]**

[1]Faculty of Pharmaceutical Sciences, Sojo University, Kumamoto, Japan
[2]School of Pharmacy, Faculty of Health Sciences, Curtin Health Innovation Research Institute, Curtin University, Perth, Western Australia, Australia
[3]DDS Research Institute, Sojo University, Kumamoto, Japan

ABSTRACT

Human serum albumin (HSA) is synthesized predominantly in the liver, and has an extraordinarily long circulatory half-life (~19 days). Since HSA is responsible for 80% of the colloidal osmotic pressure of plasma, the patients with hypoalbuminemia are typically given a plasma expander to increase the plasma volume. Clinically, using HSA as an extender is preferable to other options, such as dextran, hydroxylethyl starch, due to its long retention in plasma. However, the circulatory half-life of HSA is significantly-impacted by structural changes of HSA and pathological conditions. Recent scientific advances have revealed new information regarding the factors that influence the pharmacokinetic properties of HSA. In this chapter, we provide an overview of the impact of HSA structure and biotransformation upon the pharmacokinetic properties of HSA, and discuss new possibilities for the therapeutic potential of HSA based on its pharmacokinetic properties.

[*] Corresponding author: Masaki Otagiri, PhD, Faculty of Pharmaceutical Sciences, Sojo University, 4-22-1 Ikeda, Nishi-ku, Kumamoto 862-0082, Japan. Phone: +81-96-326-3887. Fax: +81-96-326-3887. E-mail: otagirim@ph.sojo-u.ac.jp.

1. INTRODUCTION

Human serum albumin (HSA) is synthesized predominantly in the liver, and represents the most abundant protein in plasma (approximately 5 g/dL in healthy individuals). Approximately 40% of albumin exists in the vascular space, and the remaining 60% is present in the extravascular space. HSA is mainly degraded in large organs such as muscles and skin. [1, 2] The functional capacities of HSA can be broadly classified into non-oncotic and oncotic properties. [3] Non-oncotic properties contribute to the maintenance of homeostatic balances such as antioxidation, antithrombotic and anti-inflammatory effects, and to the transport of many endogenous and exogenous substances. On the other hand, oncotic properties contribute to the modulation of fluid distribution in the various compartments of the body, because albumin in the vascular space is responsible for 70-80% of the colloidal osmotic pressure of plasma (25-33 mmHg).

Clinically, HSA is used as a plasma expander due to its oncotic properties. Patients with severe hypoalbuminemia as a result of conditions such as burns, the nephritic syndrome, chronic liver cirrhosis and hemorrhagic shock are typically treated with an infusion of albumin to correct for hypovolemia and fluid depletion in the circulation. One of the advantages of albumin infusion compared to other plasma expanders such as dextran and hydroxylethyl starch is its much longer retention in vascular space (HSA: ~19 days [4-6]; hydroxylethyl starch (130~200 kDa): 10~70 hour [7-9]; dextran (1~70 kDa): 2~42 hour (molecular weight dependent) [10, 11]; gelatin: 8~17 hour [10]). However, it is well known that the circulatory half-life of HSA is significantly impacted by structural changes of HSA and pathological conditions, which may compromise the function of the infused HSA as plasma expander. We have been investigating the effect of structural changes of HSA on its pharmacokinetic properties including circulatory half-life, biodistribution and drug-binding specificity using HSA mutants, as well as chemically and genetically modified HSA. Furthermore, novel HSA based agents that mimic the innate or acquired pharmacokinetic characteristics of HSA were developed for disease therapy and diagnosis (see "4. Therapeutic potential of HSA-based drugs with modified pharmacokinetic properties" section).

In this chapter, a brief overview of the impact of HSA structure and pathological conditions on the pharmacokinetic properties of HSA are discussed, and prospective HSA-based products that take advantage of its pharmacokinetic properties will be discussed.

2. IMPACT OF BIOTRANSFORMATIONS ON THE PHARMACOKINETIC PROPERTIES OF HSA

Albumin infusion is a standard treatment for severe hypoalbuminemia as a result of conditions such as hepatic disease and renal disease (nephritic syndrome and diabetes etc.) However, its pharmacokinetic properties are frequently influenced by the underlying disease conditions (Figure 1).

2.1. Hepatic Disease

Ohnishi et al. evaluated the pharmacokinetic profiles of native HSA in livers of cirrhosis patients without ascites or edema [12]. The half-life of native HSA in these patients was prolonged (approximately 31 day) compared to that in healthy subjects (~19 days [4-6]). In addition, albumin levels were found to fluctuate frequently in patients with ascites or edema. Remobilization of body fluids due to a third-space fluid shift into the blood stream may change the pharmacokinetic properties of the infused HSA in these patients.

Recent studies have shown that oxidative damage impairs the binding properties of HSA in advanced liver disease, and this reduced binding capacity of HSA (especially Site I and Site II ligands) is mainly related to an impaired liver function [13]. In addition, our group also showed that the binding of HSA purified from chronic liver disease patients for endogenous substances (bilirubin and tryptophan) and drugs (warfarin and diazepam) was decreased depending on the progression of the hepatic disease [14]. Furthermore, the concentration of bilirubin in plasma was increased in patients with chronic liver disease. It is known that bilirubin binds with a high affinity to a single site on HSA [15], which may result in inhibiting the binding of other ligands *via* competitive inhibition. Thus, a structural change and ligand binding displacement are part of the altered protein binding of drugs in hepatic diseases.

2.2. Renal Disease

The glomerular biological membrane allows a high filtration rate of water and small and mid-sized molecules, but does not allow larger proteins such as HSA to be filtered. These restrictions can be explained by several mechanisms, including charge repulsion in the glomerular basement membrane and barrier as a function of molecular size [16]. However, these restrictive barriers in the kidney are not viable in patients with renal disease, such as diabetes and the nephrotic syndrome, resulting in the leakage of HSA into the urine (albuminuria). In fact, animal studies have clearly shown that the plasma retention of the infused HSA decreased and this decrease was accompanied by the accumulation of HSA in the kidney and the urinary excretion of HSA in nephrotic model rats compared to healthy rats [17]. Furthermore, recent evidence indicates that the Fc receptor (FcRn) expressed in the kidney mediates albumin reuptake [18, 19], and that FcRn also extends the half-life of albumin [20]. Although domain III in HSA is required for binding to human FcRn [21, 22], the structure of HSA is altered oxidation and glycation in patients with renal disease [23-25]. In addition, the level of expression of the FcRn protein is changed in nephritic conditions [26]. These pieces of evidence suggest that the pharmacokinetic properties of infused HSA might be altered due to a high rate of glomerular filtration and a change in the reabsorption of HSA in patients with renal disease.

The binding of drugs to albumin is decreased in patients with renal diseases. One cause for this is that HSA is modified by reactive oxidative stress etc. In fact, the oxidative modification of HSA in hemodialysis patients led to alterations in the conformation of HSA, resulting in a decreased binding of site II drugs in particular to HSA [27]. Another cause is the accumulation of albumin-bound uremic toxins such as indoxyl sulfate, hippuric acid, indoleacetic acid, and 3-carboxy-4-methyl-5-propyl-2-furanpropanoic acid. These uremic toxins primarily bind to two major binding sites of HSA [28], namely sites I and II, with relatively high primary binding constants [29-31]. Thus, these uremic toxins are largely responsible for being competitive binding inhibitors that led to the impaired binding of other endogenous and exogenous substances.

Table 1. List of known pharmacokinetic properties of HSA variants with single amino-acid substitutions. Data from References [37-44]

Variants Name	Mutation	Half-life[a]	Clearance[a]		Protein binding[a]											
			Liver	Kidney	FA[b]	WF[c]	SA[c]	DAZ[c]	LA[c]	Metal[c]			hormon[c]			
										Ni^{2+}	Ca^{2+}	Zn^{2+}	PG	TS	PGF	Thy
Blenheim	1Asp→ Val	-	-	-	-	-	-	-	-	↓	↓	→	-	-	-	-
Torino	60Glu→Lys	-	-	-	↓	-	-	-	-	-	-	-	-	-	-	-
Tregasio	122Val→Glu	↓	↓	→	-	-	-	-	-	-	-	-	-	-	-	-
Hawkes Bay	177Cys→Phe	↓	↑	↑	-	-	-	-	-	-	-	-	-	-	-	-
Tradate-2	225Lys→Gln	-	-	-	↓	-	-	-	→	-	-	-	-	-	-	-
Herborn	240Lys→Glu	→	↑	→	→	-	-	-	→	-	-	-	-	-	-	-
Niigata	269 Asp→Gly	→	→	→	→	→	↓	→	→	-	-	-	→	→	↑	→
Caserta	276 Lys→Asn	↓	↑	↑	→	-	-	-	→	-	-	-	-	-	-	-
Canterbury	313 Lys→Asn	→	→	→	-	↓	↓	↓	→	-	-	-	↑	→	→	→
Brest	314Asp→Val	↓	→	→	-	-	-	-	-	-	-	-	-	-	-	-
Roma	321Glu→Lys	→	↑	↑	→	→	↓	↓	↑	-	-	-	→	↓	→	→
Sondrio	333Glu→Lys	→	→	→	-	-	-	-	→	-	-	-	-	-	-	-
Parklands	365Asp→His	→	↑	↑	-	↓	↓	→	-	-	-	-	→	→	→	→
Milano slow	375Asp→His	→	↑	→	-	-	-	-	-	-	-	-	-	-	-	-
Liprizzi	410Arg→Cys	↓	-	-	-	-	-	-	-	-	-	-	-	-	-	-
Kashmir	501Glu→Lys	→	↓	↓	→	-	-	-	→	-	-	-	-	-	-	-
Ortonovo	505Glu→Lys	-	-	-	→	-	-	-	→	-	-	-	-	-	-	-
Maku	541Lys→Glu	→	↓	↑	-	-	-	-	↓	-	-	-	-	-	-	-
Church Bay	560Lys→Glu	↓	→	↓	-	-	-	-	-	-	-	-	-	-	-	-
Paris-2	563 Asp→Asn	-	-	-	↑	-	-	-	-	-	-	-	-	-	-	-
Verona	570Glu→Lys	↑	→	→	→	→	→	→	-	-	-	-	→	→	→	→
Milano fast	573Lys→Glu	→	→	↑	-	↓	↓	→	→	-	-	-	-	-	-	-
Vanves	574 Lys→Asn	-	-	-	-	↓	↓	→	-	-	-	-	-	-	-	-

↑; increased, ↓; decreased, →; not altered, compared to each variant. -; No evaluation.

FA; fatty acid, WF; warfarin, SA; salicylate, DAZ; diazepam, LA; laurate, PG; progesterone, TS; testosterone, PGF prostaglandin $F_{2\alpha}$, Thy; L-thyroxine.

[a] Significantly change compare to corresponding normal HSA. [b] Compare the difference in total fatty acid binding between each variant and normal HSA. [c] Compare the difference in binding constant between each variant and normal HSA.

3. Impact of Structural Changes on Pharmacokinetic Properties of HSA

The pharmacokinetic properties of endogenous and infused HSA are influenced by genetic modulations and posttranslational modifications (Figure 2). The following modifications are known affectors that change the pharmacokinetic properties of HSA.

Table 2. List of known pharmacokinetic properties of chain-termination variants. Data from References [37, 40-44]

Variants Name	Mutation	Half-life[a]	Clearance[a]		Protein binding[a]							
			Liver	Kidney	FA[b]	WF[c]	SA[c]	DAZ[c]	LA[c]	Metal[c]		
										Ni^{2+}	Ca^{2+}	Zn^{2+}
Tradate	–2Arg→Cys	→	→	↑	-	-	-	-	-	-	-	-
Lille or Varese	–2Arg→His	↓	→	→	-	-	-	-	↑	↓	→	↑
Christchurch	–1Arg→ Gln	-	-	-	-	-	-	-	-	↓	↓	↓
Bazzano	567-582 substituted, 583-585 deleted	↓	↑	↑	→	-	-	-	-	-	-	-
Catania	580-582 substituted, 583-585 deleted	→	↓	↑	→	↓	→	→	→	-	-	-
Venezia	572-578 substituted, 579-585 deleted	↓	↑	↑	→	↓	↓	↓	→	-	-	-

↑; increased, ↓; decreased, →; not altered, compared to each variant. -; No evaluation.
FA; fatty acid, WF; warfarin, SA; salicylate, DAZ; diazepam, LA; laurate.
[a] Significantly change compare to corresponding normal HSA. [b] Compare the difference in total fatty acid binding between each variant and normal HSA. [c] Compare the difference in binding constant between each variant and normal HSA.

3.1. Mutations (Variant)

The advancement of genetic engineering has enabled the production of mutant proteins. Until now, our research group has designed and produced a number of recombinant single- or multiple- residue HSA mutants, using a yeast expression system, to investigate the effect of mutations on the pharmacokinetic properties of HSA. Iwao et al. evaluated the pharmacokinetic properties of single-residue mutants (H146A, K199A, W214A, R218H, R410A, Y411A), and found that the R410A mutant has the shortest half-life with increased liver clearance (Half-life: Wild-type = R218H > Y411A > H146A > K199A > W214A > R410A) [32]. In addition, Nakajou et al. reported that a recombinant triple-residue mutant (K199A/K439A/K525A) showed a decreased plasma retention with increased liver and renal clearances compared to the values for wild-type HSA [33].

There are more than 70 genetic variants of HSA [34], and more than 60 inherited variants of HSA have been identified and structurally characterized [35]. The pharmacokinetic properties of some of these genetic variants of HSA have been examined. We previously examined the pharmacokinetic profiles of 17 genetic variants of HSA with a single-residue mutation [36]. The plasma half-life of the HSA variants was different from that of native HSA (Table 1). Furthermore, our research group also investigated the half-life and uptake of three HSA genetic variants with elongated N-terminal and truncated C-terminal residues by organs [37]. It is noteworthy that one of the elongated forms (Lille) and two of the truncated forms (Bazzano and Venezia), to different degrees, had shorter half-lives compared to their corresponding normal HSA (Table 2). In addition, the albumin genetic variant Liprizzi (R410C), has also been reported to undergo rapid clearance [38]. Kragh-Hansen et al. reported that some of these HSA variants with single amino acid substitutions or with elongated N-terminal and truncated C-terminal residues, showed altered ligand binding properties [39-44] (Tables 1 and 2).

3.2. Glycation

Albumin is highly sensitive to glycation when it is exposed to high glucose concentrations in the circulation. Glycation affects the structure and function of HSA [45, 46], and eventually affects its pharmacokinetic properties. Nagai et al. showed that the half-live was the shortest for highly glycated-bovine serum albumin (BSA) < mildly glycated-BSA < native BSA [47, 48]. Furthermore, the half-life of glycated recombinant HSA (rHSA) was also shorter than that of rHSA, with increased liver and kidney uptake observed in the case of glycated rHSA [33].

Using mass spectrometry, lysine of HSA was found to be the preferred glycated site, both *in vivo* and *in vitro* [49-51]. Since the lysine residues on HSA are located near two major binding sites for drugs (site I and II) [28] were found to be glycated, the effect of glycation on drug binding at sites I and II on HSA have been studied extensively [33, 46, 52-55]. The changes in drug binding properties of glycated HSA are probably related to changes in the amount and types of modifications that occur in HSA with different degrees of glycation. Similar effect of fatty acids and other ligand binding to glycated HSA was also observed [46, 56-58].

3.3. Glycosylation

Malmö-95 (63 Asp → Asn, glycosylated at 63 Asn), Redhill (− 1 Arg retained, 320 Ala → Thr, glycosylated at 318 Asn) and Casebrook (494 Asp → Asn, glycosylated at 494 Asn) are glycosylated variants of HSA (Table 3). The pharmacokinetic of these three alloalbumins N-glycosylated in domain I (Malmö-95), II (Redhill), III (Casebrook) have been investigated by Iwao et al. [37]. The effect of glycosylation on HSA appears to depend on the position of attachment of the glycan unit. The glycosylation of 63 Asn in domain I (Malmö-95) or 494 Asn in domain III (Casebrook) has no significant effect on the half-life of these variants. In contrast, the glycosylation of 318 Asn in domain II (Redhill) results in a significant increase in the half-life of the molecule. Furthermore, the pharmacokinetic properties of a series of recombinant, mannosylated-HSA mutants (Man-rHSAs: D63N, A320T and D494N) and their triple mutant (TM-rHSA: D63N/A320T/D494N) were also evaluated by Hirata et al. [59]. They were found to be rapidly cleared from the blood circulation in the following order: TM-rHSA < D494N << A320T=D63N, consistent with their degree of mannosylation.

The effect of glycosylation on the ligand binding properties of HSA was investigated by using HSA variants isolated from bisalbuminaemic subjects. The binding of fatty acids to

Redhill and Casebrook were increased by 1.7- and 8.6-fold, respectively, while Malmö-95 carried the same amount of fatty acid, compared to their corresponding normal HSA [44]. In addition, glycosylation also caused a decrease/increase in the binding of metals and laurate to HSA (Table 3) [40, 41].

3.4. Oxidation

Anti-oxidation activity is one of the major functions of HSA, where methionine and cysteine residues on HSA accounted for 40-80% of its total antioxidant activity [60, 61]. Oxidative posttranslational modifications of HSA result in alterations in the structure, function and pharmacokinetic properties of HSA. Mildly oxidized HSA induced by a chloramine-T treatment showed the same half-life and hepatic clearance as native HSA [62]. However, extensively oxidized HSA (or advanced oxidation protein products (AOPP)) caused a shorter half-life due to an increased hepatic and splenic clearance compared to native HSA [62, 63]. The rapid clearance of oxidized HSA induced by a metal ion-catalyzed oxidation system from blood circulation was also reported [32].

Table 3. List of known pharmacokinetic properties of glycosylated variants. Data from References [37, 40-44]

Variants Name	Mutation [N-glycosylated position]	Half-life[a]	Clearance[a]		Protein binding[a]							
			Liver	Kidney	FA[b]	WF[C]	SA[C]	DAZ[C]	LA[C]	Metal[C]		
										Ni^{2+}	Ca^{2+}	Zn^{2+}
Malmö-95	63Asp→Asn [63Asn]	→	↑	→	→	-	-	-	-	-	-	-
Redhill	2Arg→Cys; 320 Ala→Thr [318Asn]	↑	↓	↓	↑	-	-	-	↓	↓	↑	↓
Casebrook	494 Asp→Asn [494Asn]	→	↑	→	↑	-	-	-	↓	-	-	-

↑; increased, ↓; decreased, →; not altered, compared to each variant. -; No evaluation

FA; fatty acid, WF; warfarin, SA; salicylate, DAZ; diazepam, LA; laurate.

[a] Significantly change compare to corresponding normal HSA. [b] Compare the difference in total fatty acid binding between each variant and normal HSA. [c] Compare the difference in binding constant between each variant and normal HSA.

The effects of HSA oxidation on drug binding have been extensively studied [64-67]. The effect of oxidation on the drug binding properties of HSA was examined by using warfarin and ketoprofen, site I and site II ligands, respectively. Except for extensive oxidation, the binding of warfarin to *in vitro* oxidized HSA was not significantly affected even after oxidation, while the binding of ketoprofen was greatly diminished [32, 62]. Oxidized HSA isolated from advanced hepatic and renal disease patients also showed an impairment in site II-ligand binding capabilities [13, 14, 27].

3.5. Others

HSA consists of three homologous domains (I-III). Matsushita et al. prepared three single recombinant domains of HSA using *Pichia pastoris*, and examined the pharmacokinetic properties of these domains [68]. In comparison to rHSA, all three domains were rapidly eliminated, with a high accumulation in the kidneys. Although the ligand binding ability of subsites Ia (located in domain II) and Ib (located in domain II) was diminished, the binding of ligands to subsite Ic (located in domain II) and site II (located in domain III) was preserved.

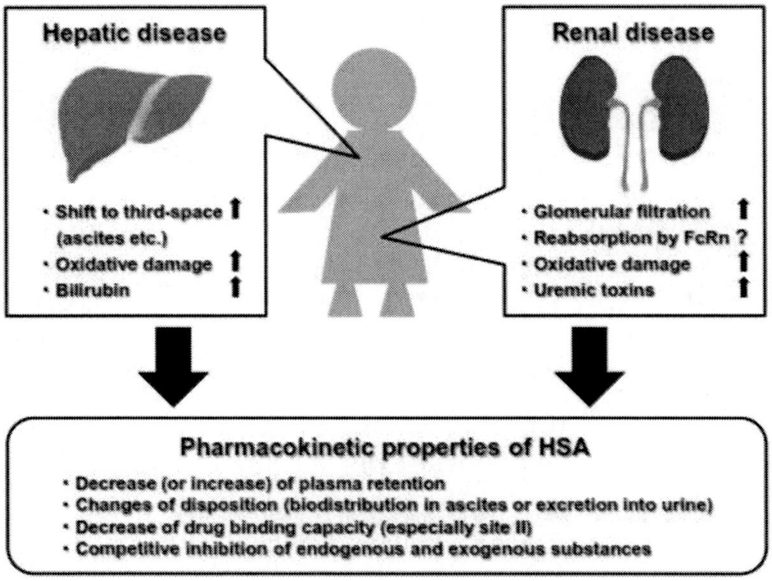

Figure 1. The effect of pathological conditions on the pharmacokinetic properties of endogenous and infused HSA.

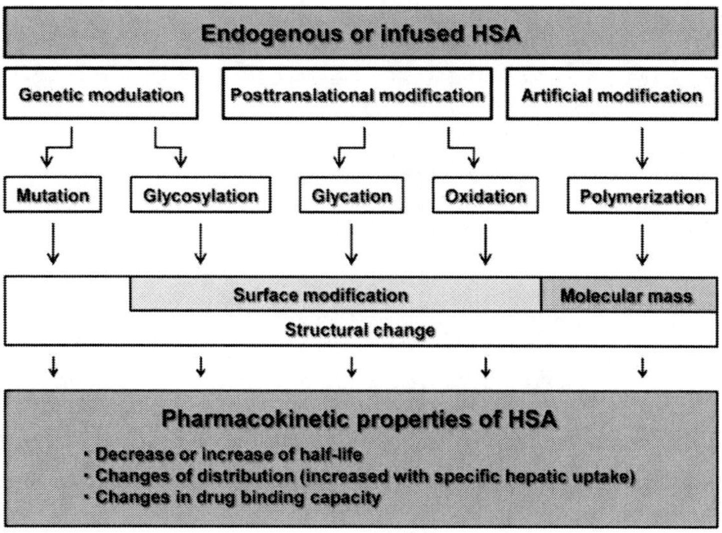

Figure 2. Scheme showing the effect of modification on the pharmacokinetic properties of endogenous and infused HSA.

On the other hand, the pharmacokinetic properties of an albumin dimer produced genetically or chemically have also been investigated by our group and by Komatsu et al., respectively [69-72]. Komatsu et al. chemically produced a rHSA dimer by linking two rHSA molecules with 1,6-bis (maleimido)hexane, which is sufficiently long (16.1 Å) to permit the Cys34 residues on the two HSA molecules to be connected without causing excessive strain on the overall structure of the protein [72, 73]. Matsushita et al. prepared a rHSA dimer in which two HSA molecules were genetically linked with an amino acid linker $(GGGGS)_2$ [69]. The drug binding capability for a site I and site II ligand for the chemical and genetic rHSA were comparable to that of the rHSA monomer [72, 74]. In addition, both the chemical and genetic rHSA dimers showed an extended plasma retention under normal conditions compared to the rHSA monomer [17, 69, 72]. Furthermore, Komatsu et al. also prepared a rHSA tetramer using *N*-succinimidyl-6-[3'-(2-pyridyldithio) propionamido] hexanoate, which connects a lysine amino group ($-NH_2$) of the rHSA core to a cysteine thiol group (-SH) of the rHSA shell [75]. The circulation half-life of the distribution phase of the rHSA tetramer was significantly longer than that of the monomer, but the half-life of the disappearance phase of the rHSA tetramer was similar to that of the monomer.

4. Therapeutic Potential of HSA-Based Drugs with Modified Pharmacokinetic Properties

There are many HSA-based drugs on the market or under clinical/ preclinical development (Figure 3). These HSA-based drugs are intended to prolong blood retention or to enhance the delivery of a drug to a target by association with HSA to acquire the pharmacokinetic characteristics of HSA. In this section, some examples of HSA-based drugs produced through different approaches that are intended to utilize the pharmacokinetic properties of HSA are discussed.

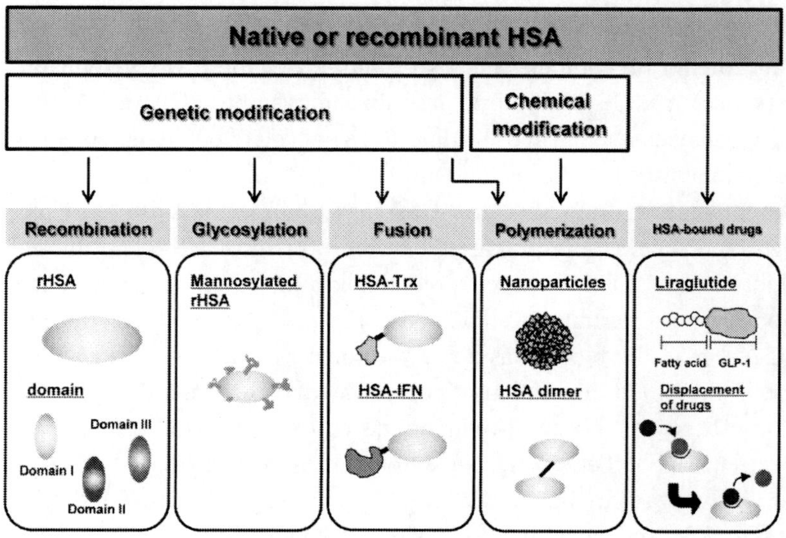

Figure 3. HSA-based drugs: recombinant HSA, glycosylated HSA, HSA fused other proteins, polymerized HSA and HSA bound drugs.

4.1. Recombinant HSA (rHSA) and Recombinant Domain

Recombinant HSA (rHSA) has been successfully produced using a methylotrophic yeast, *Pichia pastoris*. The structural and physicochemical properties of rHSA were identical to those of HSA derived from plasma [76-78]. In addition to these properties, the pharmacokinetic profiles of rHSA were deemed to be similar to plasma derived HSA in rats and dogs [79, 80]. Furthermore, Ohnishi et al. compared the pharmacokinetics of rHSA and native HSA in patients with liver cirrhosis, and concluded that the results of pharmacokinetic profiles following an intravenous administration of 25 g/day of rHSA or native HSA for 3 consecutive days indicated they were bioequivalent [12]. The binding of endogenous substances and drugs to rHSA were also within the same range as was observed for native HSA [81, 82]. These results indicate that rHSA and native HSA possess similar pharmacokinetic properties. rHSA has been approved for use as an alternative of plasma-derived HSA.

Minomo et al. identified key amino acid residues that contribute to the high affinity binding of bilirubin by a phage library and constructed an HSA mutant domain II with the objective of producing a therapeutic agent for use in the treatment of hyperbilirubinemia in patients with impaired liver function [83, 84]. They showed that the replacement of phenylalanine with arginine and arginine with leucine at positions 211 and 218, respectively, on domain II, reduced the serum bilirubin level and increased its urinary excretion in bile duct ligation model mice (hyperbilirubinemia model mice). This suggests that the rHSA mutant or the recombinant domain might be used in albumin dialysis, namely molecular adsorbents recirculatory system (MARS®), applications where the removal of bilirubin and any other protein binding toxins from the body is desirable.

4.2. HSA Fusion Proteins

The advancement of recombinant DNA technology enables two genetically fused protein molecules to be produced. In addition, as described above, the safety of rHSA produced using *Pichia pastoris* has been established, leading to the accelerated development of HSA fusion proteins. The main aim of genetic fusion with HSA is to improve the plasma retention of low molecular weight protein drugs, because HSA has a longer half-life (~19 days [4-6]). Our group designed and successfully prepared a fusion protein of HSA and thioredoxin-1(HSA-Trx) [85]. Thioredoxin-1 (Trx) has great potential as a new therapeutic protein, but its plasma half-life is too short to permit its clinical use [86-88]. As expected, the pharmacokinetic profile for HSA-Trx in the blood was about the same as that of the rHSA, and had a longer retention time in the blood circulation, compared with Trx. In addition, HSA-Trx preserved the therapeutic efficacy of Trx (anti-oxidant and anti-inflammatory effect), and appeared to show useful effects as a therapeutic agent for treating sepsis [85], lung injury [89], lung fibrosis [90], influenza virus-induced lung injury [91], contrast- and cisplatin-induced nephropathy [92, 93] and hepatitis [94]. These findings suggest that HSA-Trx is a very promising therapeutic agent for treating oxidative stress and inflammation related disorders. In other research groups, a large variety of proteins, such as peptides, cytokines, coagulation factors, enzymes, hormones and growth factors, have been genetically fused to HSA [95-99]. These fusion proteins successfully extend their retention in plasma. Therefore, protein fusion to HSA is one of the beneficial approaches for prolonging the half-life of protein drugs in the circulation.

4.3. Glycosylated rHSA

The glycosylation of protein with a carbohydrate, such as mannose, fucose and galactose, is known as one of the suitable strategies for liver-selective targeting *via* a sugar receptor. Previous studies demonstrated that chemically mannosylated-bovine serum albumin and -liposomes are largely taken up by the liver *via* a mannose receptor-mediated mechanism [100, 101]. More recently, we prepared three genetically glycosylated HSA variants (D63N/A320T/D494N) using a *Pichia pastoris* expression system [59]. This highly mannosylated-rHSA was shown to be selectively distributed in the liver, mediated by mannose receptors on Kupffer cells. Taking advantage of this liver specific distribution characteristic, we succeeded in using this highly mannosylated-rHSA as a drug carrier in the treatment of hepatic injuries, such as hepatic ischemia reperfusion injury, concanavalin A-induced hepatitis and acetaminophen-induced hepatitis [59, 102]. These results suggest that glycosylated HSA has the potential for function as a versatile carrier for liver-selective therapeutics.

4.4. Polymerization

As described in 3.5 of this chapter, both chemical and genetic rHSA dimers showed extended plasma retention profiles. Some studies have reported the possibility of using rHSA dimers in medical applications because of its pharmacokinetic properties. We previously showed that the rHSA dimer has the potential for use as a new plasma expander in various clinical situations because it showed superior blood retention characteristics even under diabetes and nephrotic conditions [17, 103]. Furthermore, the rHSA dimer is expected to have an enhanced accumulation in solid tumors *via* the enhanced permeability and retention (EPR) mechanism [104], indicating that the rHSA dimer might be used as a carrier for anticancer drugs. Ishima et al. developed a *S*-nitrosated genetic rHSA dimer for anticancer therapy to delivery nitric oxide (NO), an anticancer agent, to tumors. The accumulation of *S*-nitrosated rHSA dimer in the tumor tissue was higher than that for the *S*-nitrosated rHSA monomer and showed tumor cell death [105]. These findings suggest that the rHSA dimer has the potential for use as a biomaterial for pharmaceutical and medical applications.

HSA-based nanoparticles are ideal for encapsulating lipophilic drugs within them. Since HSA-based nanoparticles would be expected to be subject to the EPR effect when the size range is regulated at around 100 nm, it has been extensively investigated as a carrier for lipophilic anticancer drugs. In 2005, Abraxane®, an HSA paclitaxel nanoparticle, was first such drug approved in the USA and is now used worldwide. Abraxane® accumulates in the tumor through the EPR effect of solid tumors, as well as *via* a further albumin transport pathway mediated by the 60 kDa glycoprotein, gp60. Details of the anticancer mechanism of Abraxane® have been reviewed elsewhere [106-108]. To date, a number of lipophilic anticancer drugs encapsulated in HSA nanoparticles have been developed and are currently in preclinical and clinical trial stages [109-111].

Komatsu et al. developed several types of HSA-based artificial oxygen carriers (red blood cells substitute). They synthesized 2-[8-{N-(2-methylimidazolyl)}octanoyloxymethyl]-5,10,15,20-tetrakis{α,α,α,α-o-(1-methylcyclohexanamido) phenyl}porphinatoiron(II) (FecycP), and incorporated it into a rHSA dimer [72]. The rHSA-FecycP dimer solution satisfies the initial clinical requirements for an oxygen carrier as a red blood cell substitute. They also recently prepared a next generation HSA-based artificial oxygen carrier, a hemoglobin-albumin cluster (HemoAct™), in which a lysine amino group (-NH_2) of the hemoglobin core is connected to a cysteine thiol group (-SH) of the HSA shell *via* an α-succinimidyl-ε-maleimide cross-linker [112, 113]. Hemoglobin-albumin cluster possesses a moderately higher oxygen affinity than the native hemoglobin. These findings indicate that the rHSA-FecycP dimer and hemoglobin-albumin cluster have high potentials for use as red blood cell substitutes.

5.5. Drugs Bound to HSA

The glucagon-like peptide-1 (GLP-1) is a gastrointestinal hormone, and exerts important effects on several of the pathophysiological features of diabetes. GLP-1 asserts its antidiabetic effects through binding to GLP-1 receptors expressed in the pancreas and a variety of other tissues [114, 115]. Although GLP-1 has some potential as a therapeutic agent for the treatment of diabetes, the development of GLP-l as antidiabetic drug has been quite difficult due to its short half-life (~2 min). Liraglutide (Victoza®) has been developed as a new diabetic drug, and can reversibly bind to albumin through the long-chain fatty acid (palmitoyl fatty acid) that is attached *via* a spacer to an amino acid in GLP-1 [116]. Albumin protects the GLP-1 on liraglutide from metabolic degradation due to the binding of liraglutide to albumin *via* palmitoyl fatty acid, resulting in a prolonged half-life in plasma (11-15 hr) [117]. Similar to liraglutide, the half-lives of insulin detemir (Levemir®) and insulin degludec (Tresiba®) are sufficiently extend by their binding interaction to albumin *via* a long-chain fatty acid (myristyl and hexadecandioyl fatty acid for insulin detemir and insulin degludec, respectively) [116, 118].

The increment of free drug concentration due to the displacement of drugs from their albumin binding site enhances the pharmacological activity of a drug. Setoguchi et al. reported that pain relief in rheumatoid arthritis patients using a diclofenac suppository was increased by the simultaneous oral administration of nabumetone due to the competitive displacement of diclofenac from the binding site by the 6-methoxy-2-naphthylacetic acid (the active metabolite of nabumetone), because both diclofenac and the 6-methoxy-2-naphthylacetic acid bind strongly to site II on HSA [119]. In addition, a similar phenomenon was observed in a study using rats for a combination therapy involving flurbiprofen axetil and a lipid emulsion [120] or 6-methoxy-2-naphthylacetic acid [121]. These findings suggest that the drug binding properties of HSA can be exploited in drug development and further medical applications of currently well-established drugs.

CONCLUSION

HSA is clinically used as a plasma expander, and the administration of HSA is absolutely essential for treating severe hypoalbuminemia as a result of conditions such as burns, the nephritic syndrome, a reduction in the synthesis of HSA induced by chronic liver cirrhosis, and hemorrhagic shock. It is well known that posttranslational modifications of HSA and pathological conditions affect its half-life, hence the pharmacological efficacy of infused HSA. With the precise knowledge of factors that influence HSA pharmacokinetic properties, improved treatment outcomes by virtue of the proper use of HSA are feasible. On the other hand, innovative HSA preparations could be developed with the efficient utilization of the pharmacokinetic properties of chemically or genetically modified HSA.

REFERENCES

[1] Fanali G., di Masi A., Trezza V., Marino M., Fasano M., Ascenzi P. Human serum albumin: from bench to bedside. (2012). *Mol Aspects Med*. 33, (3), 209-90.

[2] Sleep D., Cameron J., Evans L. R. Albumin as a versatile platform for drug half-life extension. (2013). *Biochim Biophys Acta*. 1830, (12), 5526-34.

[3] Caraceni P., Domenicali M., Tovoli A., Napoli L., Ricci C. S., Tufoni M. et al., Clinical indications for the albumin use: still a controversial issue. (2013). *Eur J Intern Med*. 24, (8), 721-8.

[4] Takeda Y., Reeve E. B. Studies of the metabolism and distribution of albumin with autologous I131-albumin in healthy men. (1963). *J Lab Clin Med*. 61, 183-202.

[5] Bennhold H., Kallee E. Comparative studies on the half-life of I 131-labeled albumins and nonradioactive human serum albumin in a case of analbuminemia. (1959). *J Clin Invest*. 38, (5), 863-72.

[6] Beeken W. L., Volwiler W., Goldsworthy P. D., Garby L. E., Reynolds W. E., Stogsdill R. et al., Studies of I-131-albumin catabolism and distribution in normal young male adults. (1962). *J Clin Invest*. 41, 1312-33.

[7] Yamakage M., Bepperling F., Wargenau M., Miyao H. Pharmacokinetics and safety of 6 % hydroxyethyl starch 130/0.4 in healthy male volunteers of Japanese ethnicity after single infusion of 500 ml solution. (2012). *J Anesth*. 26, (6), 851-7.

[8] Waitzinger J., Bepperling F., Pabst G., Opitz J., Muller M., Francois Baron J. Pharmacokinetics and Tolerability of a New Hydroxyethyl Starch (HES) Specification [HES (130/0.4)] after Single-Dose Infusion of 6% or 10% Solutions in Healthy Volunteers. (1998). *Clin Drug Investig*. 16, (2), 151-60.

[9] Jungheinrich C., Neff T. A. Pharmacokinetics of hydroxyethyl starch. (2005). *Clin Pharmacokinet*. 44, (7), 681-99.

[10] Klotz U., Kroemer H. Clinical pharmacokinetic considerations in the use of plasma expanders. (1987). *Clin Pharmacokinet*. 12, (2), 123-35.

[11] Schwarz J. A., Koch W., Buhler V., Kaumeier S. Pharmacokinetics of low molecular (monovalent) dextran (Dx 1) in volunteers. (1981). *Int J Clin Pharmacol Ther Toxicol*. 19, (8), 358-67.

[12] Ohnishi K., Kawaguchi A., Nakajima S., Mori H., Ueshima T. A comparative pharmacokinetic study of recombinant human serum albumin with plasma-derived

human serum albumin in patients with liver cirrhosis. (2008). *J Clin Pharmacol.* 48, (2), 203-8.

[13] Oettl K., Birner-Gruenberger R., Spindelboeck W., Stueger H. P., Dorn L., Stadlbauer V. et al., Oxidative albumin damage in chronic liver failure: relation to albumin binding capacity, liver dysfunction and survival. (2013). *J Hepatol.* 59, (5), 978-83.

[14] Nagumo K., Tanaka M., Chuang V. T., Setoyama H., Watanabe H., Yamada N. et al., Cys34-cysteinylated human serum albumin is a sensitive plasma marker in oxidative stress-related chronic diseases. (2014). *PLoS One.* 9, (1), e85216.

[15] Zunszain P. A., Ghuman J., McDonagh A. F., Curry S. Crystallographic analysis of human serum albumin complexed with 4Z,15E-bilirubin-IXalpha. (2008). *J Mol Biol.* 381, (2), 394-406.

[16] Haraldsson B., Nystrom J., Deen W. M. Properties of the glomerular barrier and mechanisms of proteinuria. (2008). *Physiol Rev.* 88, (2), 451-87.

[17] Taguchi K., Urata Y., Anraku M., Watanabe H., Kawai K., Komatsu T., et al., Superior plasma retention of a cross-linked human serum albumin dimer in nephrotic rats as a new type of plasma expander. (2010). *Drug Metab Dispos.* 38, (12), 2124-9.

[18] Sarav M., Wang Y., Hack B. K., Chang A., Jensen M., Bao L. et al., Renal FcRn reclaims albumin but facilitates elimination of IgG. (2009). *J Am Soc Nephrol.* 20, (9), 1941-52.

[19] Kim J., Hayton W. L., Robinson J. M., Anderson C. L. Kinetics of FcRn-mediated recycling of IgG and albumin in human: pathophysiology and therapeutic implications using a simplified mechanism-based model. (2007). *Clin Immunol.* 122, (2), 146-55.

[20] Chaudhury C., Mehnaz S., Robinson J. M., Hayton W. L., Pearl D. K., Roopenian D. C. et al., The major histocompatibility complex-related Fc receptor for IgG (FcRn) binds albumin and prolongs its lifespan. (2003). *J Exp Med.* 197, (3), 315-22.

[21] Chaudhury C., Brooks C. L., Carter D. C., Robinson J. M., Anderson C. L. Albumin binding to FcRn: distinct from the FcRn-IgG interaction. (2006). *Biochemistry.* 45, (15), 4983-90.

[22] Andersen J. T., Dalhus B., Cameron J., Daba M. B., Plumridge A., Evans L. et al., Structure-based mutagenesis reveals the albumin-binding site of the neonatal Fc receptor. (2012). *Nat Commun.* 3, 610.

[23] Cao H., Chen T., Shi Y. Glycation of human serum albumin in diabetes: impacts on the structure and function. (2015). *Curr Med Chem.* 22,(1), 4-13.

[24] Monacelli F., Storace D., D'Arrigo C., Sanguineti R., Borghi R., Pacini D. et al., Structural alterations of human serum albumin caused by glycative and oxidative stressors revealed by circular dichroism analysis. (2013). *Int J Mol Sci.* 14, (6), 10694-709.

[25] Jurasekova Z., Tinti A., Torreggiani A. Use of Raman spectroscopy for the identification of radical-mediated damages in human serum albumin. (2011). *Anal Bioanal Chem.* 400, (9), 2921-31.

[26] Feng S. T., Gan H. L., Sun J. Y., Jiang T., Liu B. L., Zhao Z. H. et al., Expression of neonatal Fc receptor on human nephritis and rat nephritis models. (2012). *Zhonghua Bing Li Xue Za Zhi.* 41, (2), 81-5.

[27] Mera K., Anraku M., Kitamura K., Nakajou K., Maruyama T., Otagiri M. The structure and function of oxidized albumin in hemodialysis patients: Its role in elevated oxidative stress via neutrophil burst. (2005). *Biochem Biophys Res Commun.* 334, (4), 1322-8.
[28] Curry S., Mandelkow H., Brick P., Franks N. Crystal structure of human serum albumin complexed with fatty acid reveals an asymmetric distribution of binding sites. (1998). *Nat Struct Biol.* 5, (9), 827-35.
[29] Watanabe H., Miyamoto Y., Otagiri M., Maruyama T. Update on the pharmacokinetics and redox properties of protein-bound uremic toxins. (2011). *J Pharm Sci.* 100, (9), 3682-95.
[30] Sakai T., Takadate A., Otagiri M. Characterization of binding site of uremic toxins on human serum albumin. (1995). *Biol Pharm Bull.* 18, (12), 1755-61.
[31] Ghuman J., Zunszain P. A., Petitpas I., Bhattacharya A. A., Otagiri M., Curry S. Structural basis of the drug-binding specificity of human serum albumin. (2005). *J Mol Biol.* 353, (1), 38-52.
[32] Iwao Y., Anraku M., Yamasaki K., Kragh-Hansen U., Kawai K., Maruyama T. et al., Oxidation of Arg-410 promotes the elimination of human serum albumin. (2006). *Biochim Biophys Acta.* 1764, (4), 743-9.
[33] Nakajou K., Watanabe H., Kragh-Hansen U., Maruyama T., Otagiri M. The effect of glycation on the structure, function and biological fate of human serum albumin as revealed by recombinant mutants. (2003). *Biochim Biophys Acta.* 1623, (2-3), 88-97.
[34] Peters T. *Welcome to the Albumin Website* 2015 February 12. Available from: http://albumin.org/.
[35] Minchiotti L., Campagnoli M., Rossi A., Cosulich M, E., Monti M., Pucci P. et al., A nucleotide insertion and frameshift cause albumin Kenitra, an extended and O-glycosylated mutant of human serum albumin with two additional disulfide bridges. (2001). *Eur J Biochem.* 268, (2), 344-52.
[36] Iwao Y., Hiraike M., Kragh-Hansen U., Mera K., Noguchi T., Anraku M. et al., Changes of net charge and alpha-helical content affect the pharmacokinetic properties of human serum albumin. (2007). *Biochim Biophys Acta.* 1774, (12), 1582-90.
[37] Iwao Y., Hiraike M., Kragh-Hansen U., Kawai K., Suenaga A., Maruyama T. et al., Altered chain-length and glycosylation modify the pharmacokinetics of human serum albumin. (2009). *Biochim Biophys Acta.* 1794, (4), 634-41.
[38] Ishima Y., Sawa T., Kragh-Hansen U., Miyamoto Y., Matsushita S., Akaike T. et al., S-Nitrosylation of human variant albumin Liprizzi (R410C) confers potent antibacterial and cytoprotective properties. (2007). *J Pharmacol Exp Ther.* 320, (3), 969-77.
[39] Kragh-Hansen U., Minchiotti L., Brennan S. O., Sugita O. Hormone binding to natural mutants of human serum albumin. (1990). *Eur J Biochem.* 193, (1), 169-74.
[40] Kragh-Hansen U., Brennan S. O., Minchiotti L., Galliano M. Modified high-affinity binding of Ni^{2+}, Ca^{2+} and Zn^{2+} to natural mutants of human serum albumin and proalbumin. (1994). *Biochem J.* 301 (Pt 1), 217-23.
[41] Kragh-Hansen U., Pedersen A. O., Galliano M., Minchiotti L., Brennan S. O., Tarnoky A. L. et al., High-affinity binding of laurate to naturally occurring mutants of human serum albumin and proalbumin. (1996). *Biochem J.* 320 (Pt 3), 911-6.
[42] Kragh-Hansen U., Brennan S. O., Galliano M., Sugita O. Binding of warfarin, salicylate, and diazepam to genetic variants of human serum albumin with known mutations. (1990). *Mol Pharmacol.* 37, (2), 238-42.

[43] Vestberg K., Galliano M., Minchiotti L., Kragh-Hansen U. High-affinity binding of warfarin, salicylate and diazepam to natural mutants of human serum albumin modified in the C-terminal end. (1992). *Biochem Pharmacol.* 44, (8), 1515-21.

[44] Kragh-Hansen U., Nielsen H., Pedersen AO. In vivo and *in vitro* binding of fatty acids to genetic variants of human serum albumin. (1995). *Int J Clin Pharmacol Res.* 15, (2), 33-43.

[45] Rondeau P., Bourdon E. The glycation of albumin: structural and functional impacts. (2011). *Biochimie.* 93, (4), 645-58.

[46] Anguizola J., Matsuda R., Barnaby O. S., Hoy K. S., Wa C., DeBolt E. et al., Review: Glycation of human serum albumin. (2013). *Clin Chim Acta.* 425, 64-76.

[47] Nagai R., Mera K., Fujiwara Y., Nagai M., Otagiri M. Comparison of pharmacokinetics between highly and mildly modified AGE proteins in mice. (2008). *Ann N Y Acad Sci.* 1126, 325-7.

[48] Nagai R., Mera K., Nakajou K., Fujiwara Y., Iwao Y., Imai H. et al., The ligand activity of AGE-proteins to scavenger receptors is dependent on their rate of modification by AGEs. (2007). *Biochim Biophys Acta.* 1772, (11-12), 1192-8.

[49] Priego-Capote F., Scherl A., Muller M., Waridel P., Lisacek F., Sanchez J. C. Glycation isotopic labeling with 13C-reducing sugars for quantitative analysis of glycated proteins in human plasma. (2010). *Mol Cell Proteomics.* 9, (3), 579-92.

[50] Stefanowicz P., Kijewska M., Kluczyk A., Szewczuk Z. Detection of glycation sites in proteins by high-resolution mass spectrometry combined with isotopic labeling. (2010). *Anal Biochem.* 400, (2), 237-43.

[51] Barnaby O. S., Cerny R. L., Clarke W., Hage DS. Comparison of modification sites formed on human serum albumin at various stages of glycation. (2011). *Clin Chim Acta.* 412, (3-4), 277-85.

[52] Joseph K. S., Hage DS. The effects of glycation on the binding of human serum albumin to warfarin and L-tryptophan. (2010). *J Pharm Biomed Anal.* 53, (3), 811-8.

[53] Okabe N., Hashizume N. Drug binding properties of glycosylated human serum albumin as measured by fluorescence and circular dichroism. (1994). *Biol Pharm Bull* 17, (1), 16-21.

[54] Jackson A. J., Anguizola J., Pfaunmiller E. L., Hage D. S. Use of entrapment and high-performance affinity chromatography to compare the binding of drugs and site-specific probes with normal and glycated human serum albumin. (2013). *Anal Bioanal Chem.* 405, (17), 5833-41.

[55] Matsuda R., Anguizola J., Joseph K. S., Hage D. S. Analysis of drug interactions with modified proteins by high-performance affinity chromatography: binding of glibenclamide to normal and glycated human serum albumin. (2012). *J Chromatogr A.* 1265, 114-22.

[56] Baraka-Vidot J., Guerin-Dubourg A., Bourdon E., Rondeau P. Impaired drug-binding capacities of *in vitro* and *in vivo* glycated albumin. (2012). *Biochimie.* 94, (9), 1960-7.

[57] Koizumi K., Ikeda C., Ito M., Suzuki J., Kinoshita T., Yasukawa K. et al., Influence of glycosylation on the drug binding of human serum albumin. (1998). *Biomed Chromatogr.* 12, (4), 203-10.

[58] Basiaga S. B., Hage D. S. Chromatographic studies of changes in binding of sulfonylurea drugs to human serum albumin due to glycation and fatty acids. (2010). *J Chromatogr B Analyt Technol Biomed Life Sci*. 878, (30), 3193-7.

[59] Hirata K., Maruyama T., Watanabe H., Maeda H., Nakajou K., Iwao Y. et al., Genetically engineered mannosylated-human serum albumin as a versatile carrier for liver-selective therapeutics. (2010). *J Control Release*. 145, (1), 9-16.

[60] Bourdon E., Loreau N., Lagrost L., Blache D. Differential effects of cysteine and methionine residues in the antioxidant activity of human serum albumin. (2005). *Free Radic Res*. 39, (1), 15-20.

[61] Iwao Y., Ishima Y., Yamada J., Noguchi T., Kragh-Hansen U., Mera K. et al., Quantitative evaluation of the role of cysteine and methionine residues in the antioxidant activity of human serum albumin using recombinant mutants. (2012). *IUBMB Life*. 64, (5), 450-4.

[62] Anraku M., Kragh-Hansen U., Kawai K., Maruyama T., Yamasaki Y., Takakura Y. et al., Validation of the chloramine-T induced oxidation of human serum albumin as a model for oxidative damage in vivo. (2003). *Pharm Res*. 20, (4), 684-92.

[63] Iwao Y., Anraku M., Hiraike M., Kawai K., Nakajou K., Kai T. et al., The structural and pharmacokinetic properties of oxidized human serum albumin, advanced oxidation protein products (AOPP). (2006). *Drug Metab Pharmacokinet*. 21, (2), 140-6.

[64] Maciazek-Jurczyk M., Sulkowska A., Rownicka-Zubik J. Alteration of methotrexate binding to human serum albumin induced by oxidative stress. Spectroscopic comparative study. (2015). *Spectrochim Acta A Mol Biomol Spectrosc*. doi: 10.1016/j.saa.2014.12.113.

[65] Oettl K., Stauber R. E. Physiological and pathological changes in the redox state of human serum albumin critically influence its binding properties. (2007). *Br J Pharmacol*. 151, (5), 580-90.

[66] Anraku M., Yamasaki K., Maruyama T., Kragh-Hansen U., Otagiri M. Effect of oxidative stress on the structure and function of human serum albumin. (2001). *Pharm Res*. 18, (5), 632-9.

[67] Maciazek-Jurczyk M., Sulkowska A. Spectroscopic analysis of the impact of oxidative stress on the structure of human serum albumin (HSA) in terms of its binding properties. (2015). *Spectrochim Acta A Mol Biomol Spectrosc*. 136 Pt B, 265-82.

[68] Matsushita S., Isima Y., Chuang V. T., Watanabe H., Tanase S., Maruyama T. et al., Functional analysis of recombinant human serum albumin domains for pharmaceutical applications. (2004). *Pharm Res*. 21, (10), 1924-32.

[69] Matsushita S., Chuang V. T., Kanazawa M., Tanase S., Kawai K., Maruyama T. et al., Recombinant human serum albumin dimer has high blood circulation activity and low vascular permeability in comparison with native human serum albumin. (2006). *Pharm Res*. 23, (5), 882-91.

[70] Edsall J., T., Maybury R., H., Simpson R. B., Straessle R. Dimerization of Serum Mercaptalbumin in Presence of Mercurials. II. Studies with a Bifunctional Organic Mercurial. (1954). *J Am Chem Soc*. 76, 3131-38.

[71] Straessle R. A Disulfide Dimer of Human Mercaptalbumin1a,b. (1954). *J Am Chem Soc*. 76, 3138-42.

[72] Komatsu T., Oguro Y., Teramura Y., Takeoka S., Okai J., Anraku M. et al., Physicochemical characterization of cross-linked human serum albumin dimer and its

synthetic heme hybrid as an oxygen carrier. (2004). *Biochim Biophys Acta*. 1675, (1-3), 21-31.

[73] Komatsu T., Hamamatsu K., Tsuchida E. Cross-Linked Human Serum Albumin Dimer Incorporating Sixteen (Tetraphenylporphinato)iron(II) Derivatives: Synthesis, Characterization, and O2-Binding Property. (1999). *Macromolecules*. 32, (25), 8388-91.

[74] Taguchi K., Chuang V. T., Maruyama T., Otagiri M. Pharmaceutical aspects of the recombinant human serum albumin dimer: structural characteristics, biological properties, and medical applications. (2012). *J Pharm Sci*. 101, (9), 3033-46.

[75] Komatsu T., Oguro Y., Nakagawa A., Tsuchida E. Albumin clusters: structurally defined protein tetramer and oxygen carrier including thirty-two iron(II) porphyrins. (2005). *Biomacromolecules*. 6, (6), 3397-403.

[76] Ohtani W., Masaki A., Ikeda Y., Hirose M., Chuganji M., Takeshima K. et al , Structure of recombinant human serum albumin from Pichia pastoris. (1997). *Yakugaku Zasshi*. 117, (4), 220-32.

[77] Sugio S., Kashima A., Mochizuki S., Noda M., Kobayashi K. Crystal structure of human serum albumin at 2.5 A resolution. (1999). *Protein Eng*. 12, (6), 439-46.

[78] Ohtani W., Nawa Y., Takeshima K., Kamuro H., Kobayashi K., Ohmura T. Physicochemical and immunochemical properties of recombinant human serum albumin from Pichia pastoris. (1998). *Anal Biochem*. 256, (1), 56-62.

[79] Okano K., Ohkubo M., Sogame Y., Haruta T., Tanaka M., Nogami T. et al., Metabolic fate of recombinant human serum albumin (rHSA) (1) - Plasma concentration after a single or repeated intravenous administration of ^{125}I-rHSA to rats and dogs. (1997). *Jan Pharmacol Ther*. 25, (SUPPL. 8), 197-205.

[80] Okano K., Sogame Y., Ohkubo M., Haruta T., Tsuda Y. Metabolic fate of recombinant human serum albumin (rHSA) (2) - Distribution, metabolism and excretion of rHSA after a single or repeated intravenous administration of ^{125}I-rHSA to rats and excretion after a single intravenous administration of ^{125}I-rHSA to dogs. (1997). *Jan Pharmacol Ther*. 25, (SUPPL. 8), 207-18.

[81] Chuang V. T., Otagiri M. Recombinant human serum albumin. (2007). *Drugs Today (Barc)*. 43, (8), 547-61.

[82] Takeshima K., Ohtani W., Sumi A, Ohmura T. Ligand binding properties and esterase-like activity of recombinant human serum albumin. (1996). *Yakugaku Zasshi*. 116, (8) 622-9.

[83] Minomo A., Ishima Y., Chuang V. T., Suwa Y., Kragh-Hansen U., Narisoko T. et al., Albumin domain II mutant with high bilirubin binding affinity has a great potential as serum bilirubin excretion enhancer for hyperbilirubinemia treatment. (2013). *Biochim Biophys Acta*. 1830, (4), 2917-23.

[84] Minomo A., Ishima Y., Kragh-Hansen U., Chuang V. T., Uchida M., Taguchi K. et al., Biological characteristics of two lysines on human serum albumin in the high-affinity binding of 4Z,15Z-bilirubin-IXalpha revealed by phage display. (2011). *FEBS J*. 278, (21), 4100-11.

[85] Ikuta S., Chuang V. T., Ishima Y., Nakajou K., Furukawa M., Watanabe H. et al., Albumin fusion of thioredoxin--the production and evaluation of its biological activity for potential therapeutic applications. (2010). *J Control Release*. 147, (1), 17-23.

[86] Nakamura H., Hoshino Y., Okuyama H., Matsuo Y., Yodoi J. Thioredoxin 1 delivery as new therapeutics. (2009). *Adv Drug Deliv Rev*. 61, (4), 303-9.

[87] Yoshihara E., Masaki S., Matsuo Y., Chen Z., Tian H., Yodoi J. Thioredoxin/Txnip: redoxisome, as a redox switch for the pathogenesis of diseases. (2014). *Front Immunol*. 4, 514.

[88] Nakamura H., Herzenberg L. A., Bai J., Araya S., Kondo N., Nishinaka Y. et al., Circulating thioredoxin suppresses lipopolysaccharide-induced neutrophil chemotaxis. (2001). *Proc Natl Acad Sci U S A*. 98, (26), 15143-8.

[89] Furukawa M., Tanaka R., Chuang V. T., Ishima Y., Taguchi K., Watanabe H. et al., Human serum albumin-thioredoxin fusion protein with long blood retention property is effective in suppressing lung injury. (2011). *J Control Release*. 154, (2), 189-95.

[90] Tanaka R., Watanabe H., Kodama A., Chuang V. T., Ishima Y., Hamasaki K. et al., Long-acting human serum albumin-thioredoxin fusion protein suppresses bleomycin-induced pulmonary fibrosis progression. (2013). *J Pharmacol Exp Ther*. 345, (2), 271-83.

[91] Tanaka R., Ishima Y., Enoki Y., Kimachi K., Shirai T., Watanabe H. et al., Therapeutic impact of human serum albumin-thioredoxin fusion protein on influenza virus-induced lung injury mice. (2014). *Front Immunol*. 5, 561.

[92] Kodama A., Watanabe H., Tanaka R., Tanaka H., Chuang V. T., Miyamoto Y. et al., A human serum albumin-thioredoxin fusion protein prevents experimental contrast-induced nephropathy. (2013). *Kidney Int*. 83, (3), 446-54.

[93] Kodama A., Watanabe H., Tanaka R., Kondo M., Chuang V. T., Wu Q. et al., Albumin fusion renders thioredoxin an effective anti-oxidative and anti-inflammatory agent for preventing cisplatin-induced nephrotoxicity. (2014). *Biochim Biophys Acta*. 1840, (3), 1152-62.

[94] Tanaka R., Ishima Y., Maeda H., Kodama A., Nagao S., Watanabe H. et al., Albumin fusion prolongs the antioxidant and anti-inflammatory activities of thioredoxin in mice with acetaminophen-induced hepatitis. (2014). *Mol Pharm*. 11, (4), 1228-38.

[95] Wang W., Ou Y., Shi Y. AlbuBNP, a recombinant B-type natriuretic peptide and human serum albumin fusion hormone, as a long-term therapy of congestive heart failure. (2004). *Pharm Res*. 21, (11), 2105-11.

[96] Tomkin G. H. Albiglutide, an albumin-based fusion of glucagon-like peptide 1 for the potential treatment of type 2 diabetes. (2009). *Curr Opin Mol Ther*. 11, (5), 579-88.

[97] Santagostino E., Negrier C., Klamroth R., Tiede A., Pabinger-Fasching I., Voigt C. et al., Safety and pharmacokinetics of a novel recombinant fusion protein linking coagulation factor IX with albumin (rIX-FP) in hemophilia B patients. (2012). *Blood*. 120, (12), 2405-11.

[98] Subramanian G. M., Fiscella M., Lamouse-Smith A., Zeuzem S., McHutchison J. G. Albinterferon alpha-2b: a genetic fusion protein for the treatment of chronic hepatitis C. (2007). *Nat Biotechnol*. 25, (12), 1411-9.

[99] Osborn B. L., Sekut L., Corcoran M., Poortman C., Sturm B., Chen G. et al., Albutropin: a growth hormone-albumin fusion with improved pharmacokinetics and pharmacodynamics in rats and monkeys. (2002). *Eur J Pharmacol*. 456, (1-3), 149-58.

[100] Higuchi Y., Nishikawa M., Kawakami S., Yamashita F., Hashida M. Uptake characteristics of mannosylated and fucosylated bovine serum albumin in primary

cultured rat sinusoidal endothelial cells and Kupffer cells. (2004). *Int J Pharm*. 287, (1-2), 147-54.

[101] Opanasopit P., Sakai M., Nishikawa M., Kawakami S., Yamashita F., Hashida M. Inhibition of liver metastasis by targeting of immunomodu- lators using mannosylated liposome carriers. (2002). *J Control Release*. 80, (1-3), 283-94.

[102] Maeda H., Hirata K., Watanabe H., Ishima Y., Chuang V. T., Taguchi K. et al., Polythiol-containing, recombinant mannosylated-albumin is a superior CD68+/CD206+ Kupffer cell-targeted nanoantioxidant for treatment of two acute hepatitis models. (2015). *J Pharmacol Exp Ther*. 352, (2), 244-57.

[103] Taguchi K., Chuang V. T., Yamasaki K., Urata Y., Tanaka R., Anraku M. et al., Cross-linked human serum albumin dimer has the potential for use as a plasma-retaining agent for the fatty acid-conjugated antidiabetic drugs. (2015). *J Pharm Pharmacol*. 67, (2), 255-63.

[104] Maeda H., Wu J., Sawa T., Matsumura Y., Hori K. Tumor vascular permeability and the EPR effect in macromolecular therapeutics: a review. (2000). *J Control Release*. 65, (1-2), 271-84.

[105] Ishima Y., Chen D., Fang J., Maeda H., Minomo A., Kragh-Hansen U. et al., S-Nitrosated human serum albumin dimer is not only a novel anti-tumor drug but also a potentiator for anti-tumor drugs with augmented EPR effects. (2012). *Bioconjug Chem*. 23, (2), 264-71.

[106] Kratz F., Elsadek B. Clinical impact of serum proteins on drug delivery. (2012). *J Control Release*. 161, (2), 429-45.

[107] Kratz F. Albumin as a drug carrier: design of prodrugs, drug conjugates and nanoparticles. (2008). *J Control Release*. 132, (3), 171-83.

[108] Fu Q., Sun J., Zhang W., Sui X., Yan Z., He Z. Nanoparticle albumin-bound (NAB) technology is a promising method for anti-cancer drug delivery. (2009). *Recent Pat Anticancer Drug Discov*. 4, (3), 262-72.

[109] Jiang S., Gong X., Zhao X., Zu Y. Preparation, characterization, and antitumor activities of folate-decorated docetaxel-loaded human serum albumin nanoparticles. (2015). *Drug Deliv*. 22, (2), 206-13.

[110] Gonzalez-Angulo A. M., Meric-Bernstam F., Chawla S., Falchook G., Hong D., Akcakanat A. et al., Weekly nab-Rapamycin in patients with advanced nonhematologic malignancies: final results of a phase I trial. (2013). *Clin Cancer Res*. 19, (19), 5474-84

[111] Tang Q. S., Chen D. Z., Xue W. Q., Xiang J. Y., Gong Y. C., Zhang L. et al., Preparation and biodistribution of 188Re-labeled folate conjugated human serum albumin magnetic cisplatin nanoparticles (188Re-folate-CDDP/HSA MNPs) in vivo. (2011). *Int J Nanomedicine*. 6, 3077-85.

[112] Tomita D., Kimura T., Hosaka H., Daijima Y., Haruki R., Ludwig K. et al., Covalent core-shell architecture of hemoglobin and human serum albumin as an artificial O2 carrier. (2013). *Biomacromolecules*. 14, (6), 1816-25.

[113] Hosaka H., Haruki R., Yamada K., Bottcher C., Komatsu T. Hemoglobin-albumin cluster incorporating a Pt nanoparticle: artificial O2 carrier with antioxidant activities. (2014). *PLoS One*. 9, (10), e110541.

[114] Pyke C., Heller R. S., Kirk R. K., Orskov C., Reedtz-Runge S., Kaastrup P. et al., GLP-1 receptor localization in monkey and human tissue: novel distribution revealed with extensively validated monoclonal antibody. (2014). *Endocrinology*. 155, (4), 1280-90.

[115] Lund A., Knop F. K., Vilsboll T. Glucagon-like peptide-1 receptor agonists for the treatment of type 2 diabetes: differences and similarities. (2014). *Eur J Intern Med*. 25, (5), 407-14.

[116] Ahren B., Burke B. Using albumin to improve the therapeutic properties of diabetes treatments. (2012). *Diabetes Obes Metab*. 14, (2), 121-9.

[117] Degn K. B., Juhl C. B., Sturis J., Jakobsen G., Brock B., Chandramouli V. et al., One week's treatment with the long-acting glucagon-like peptide 1 derivative liraglutide (NN2211) markedly improves 24-h glycemia and alpha- and beta-cell function and reduces endogenous glucose release in patients with type 2 diabetes. (2004). *Diabetes*. 53, (5), 1187-94.

[118] Gough S. C., Harris S., Woo V., Davies M. Insulin degludec: overview of a novel ultra long-acting basal insulin. (2013). *Diabetes Obes Metab*. 15, (4), 301-9.

[119] Setoguchi N., Takamura N., Fujita K., Ogata K., Tokunaga J., Nishio T. et al., A diclofenac suppository-nabumetone combination therapy for arthritic pain relief and a monitoring method for the diclofenac binding capacity of HSA site II in rheumatoid arthritis. (2013). *Biopharm Drug Dispos*. 34, (2), 125-36.

[120] Osaki T., Ozaki M., Takamura N., Ogata K., Tokunaga J., Setoguchi N. et al., Albumin-binding of diclofenac and the effect of a site II inhibitor in the aqueous humor of cataract patients with the instillation of diclofenac. (2014). *Biopharm Drug Dispos*. 35, (4), 218-27.

[121] Ogata K., Takamura N., Tokunaga J., Ikeda T., Setoguchi N., Tanda K. et al., A novel injection strategy of flurbiprofen axetil by inhibiting protein binding with 6-methoxy-2-naphthylacetic acid. (2014). *Eur J Drug Metab Pharmacokinet*. doi: 10.1007/s13318-014-0248-z

In: Human Serum Albumin (HSA)
Editor: Travis Stokes

ISBN: 978-1-63482-963-2

Chapter 5

GLYCATIVE AND OXIDATIVE MODIFICATION OF HUMAN SERUM ALBUMIN: CONFORMATIONAL AND FUNCTIONAL CHANGES

Makoto Anraku,*[1,4]*, Ulrich Kragh-Hansen*[2]*,
***Toru Maruyama*[3] *and Masaki Otagiri*[1,4,*]**
[1]Faculty of Pharmaceutical Sciences, Sojo University, Kumamoto, Japan
[2]Department of Biomedicine, University of Aarhus, Denmark
[3]Graduate School of Pharmaceutical Sciences,
Kumamoto University, Kumamoto, Japan
[4]DDS Research Institute, Sojo University, Kumamoto, Japan

ABSTRACT

Glycative and oxidative stress are frequently observed in numerous disease states. From recent studies, it has become evident that protein glycation and oxidation play major roles in the activity, unfolding, and degradation of proteins, as well as for cell function. Human serum albumin (HSA), the major protein in the blood circulation, frequently undergoes increased glycation and oxidation in major diseases such as diabetes and renal failure. This chapter examines the different aspects of albumin glycation and oxidation by focusing on the impact of these processes on the structural, biological and physiological modification of the protein. Further, the importance of glycated and oxidized albumin as a biological marker for many diseases such as diabetes and renal failure is discussed.

[*] Corresponding author at: Faculty of Pharmaceutical Sciences, Sojo University, Nishi-ku, Kumamoto, Japan, Tel.: +81-96-326-3887, fax: +81-96-326-5048, E-mail address: otagirim@ph.sojo-u.ac.jp (M. Otagiri).

1. INTRODUCTION

Over time, body proteins undergo non-enzymatic post-translational modifications, which include glycation and oxidation. Such alterations can affect our tissues and their levels accumulate throughout our lifetime with an undeniable effect on metabolism. These molecular and cellular alterations, although initially harmless, can become damaging and pathogenic when they reach high levels. Among the various targets that are sensitive to alteration, it has been suggested that blood plasma is sustainably and continually exposed to numerous metabolites, which could, in particular, induce oxidative stress.

Human serum albumin (HSA) is the most abundant plasma protein in the circulation, with normal concentrations in the range of 35-50 g/L. This protein possesses a wide range of biological properties [1, 2], and there is ample evidence suggesting that it exerts significant antioxidant activity [3]. In fact, albumin represents the major and predominant circulating antioxidant molecule in the plasma is exposed to continuous stress [4-6]. In the case of diabetes, glycative stress could originate from various processes, such as the excessive production of oxygen radicals from the autoxidation of glucose, glycoxidized proteins, and the glycoxidation of antioxidative proteins, with albumin being the most important of these [7]. Compiled data from biochemical, animal and epidemiological studies strongly support the hypothesis that glycoxidative modifications of circulating HSA play a pivotal and causative role in the pathogenesis of diabetes [8-11]. In the case of chronic renal failure (CRF), oxidative stress is enhanced in correlation with the level of renal dysfunction [13]. HSA has been reported to undergo massive oxidation in cases of CRF [14]. Further, Cys-34 of HSA is highly accessible to reactive free radicals such as H_2O_2 [15] and carbon-centered free radicals [16], as well as other oxidizing agents, such as nitric oxide (NO) [17-19] and peroxynitrite ($ONOO^-$) [15, 19]. Thus, the thiol-dependent antioxidant effect of HSA has recently attracted interest as a new function and it has been proposed that HSA represents a very abundant and potentially important circulating antioxidant [3, 20].

This chapter reviews the different aspects of HSA glycation and oxidation by focusing on the impact of these processes on the structural, biological and physiological properties of the protein. The effects of *in vitro* and *in vivo* glycation and oxidation on the binding properties of HSA, in particular, are discussed. Finally, and with regard to its clinical and biological relevance, the importance of glycated or oxidized HSA as biological markers of many diseases such as diabetes and CRF is discussed.

2. NON-ENZYMATIC GLYCATION OF HSA

Non-enzymatic glycation is one of the underlying factors contributing to the development of diabetic complications. HSA is non-enzymatically glycated by reducing sugars that are present in the general circulation. The reference range of glycated HSA in normal humans is 6-10%. However, this proportion typically increases to between 20% and 30% in hyperglycemic patients. This section focuses on the different aspects of HSA glycation by focusing on the impact of this process on structural, biological and physiological modifications of the protein in *in vitro* and *in vivo* studies.

2.1. Structural Analysis of Glycated HSA

Numerous studies have identified the main sites in HSA that are available for modification *in vivo*. The higher accessibility of some amino acid residues depends on the tertiary structure conformation of HSA. Because of their high nucleophilic properties, lysine and arginine are the sole residues prone to glycation. About 29 such sites have been described, including 18 lysine residues, as shown in Figure 1. In several studies, lysine-525 is reported to be the predominant site for the non-enzymatic glycosylation of HSA *in vivo* [21-23]. Lys-525 appears to be the most reactive glycation site, despite its low surface exposure and accessibility in the native conformation [24]. Along with Lys-525, three other lysine residues at positions 199, 281, and 439, have been clearly established, but with a lower contribution to the overall glycation than Lys-525 [25-27]. On the other hand, Arg-410 was found to be a hotspot for modification by methylglyoxal in HSA *in vivo*. Minor arginine sites that are involved in glycation, such as Arg-114, Arg-160, Arg-186, Arg-218 and Arg-428, have also been identified [27, 28]. Nε-carboxymethyl-arginine (CMA) constitutes an acid labile AGE implying arginine residues. CMA represents the major glycated product in collagen and has been detected in small amounts in HSA obtained from diabetic patients [28, 29]. Further, AGE formation in the case of HSA is associated with extensive damage to histidine and tryptophan side chains, main chain fragmentation, and the loss of both secondary and tertiary structure as a result of oxidation [30].

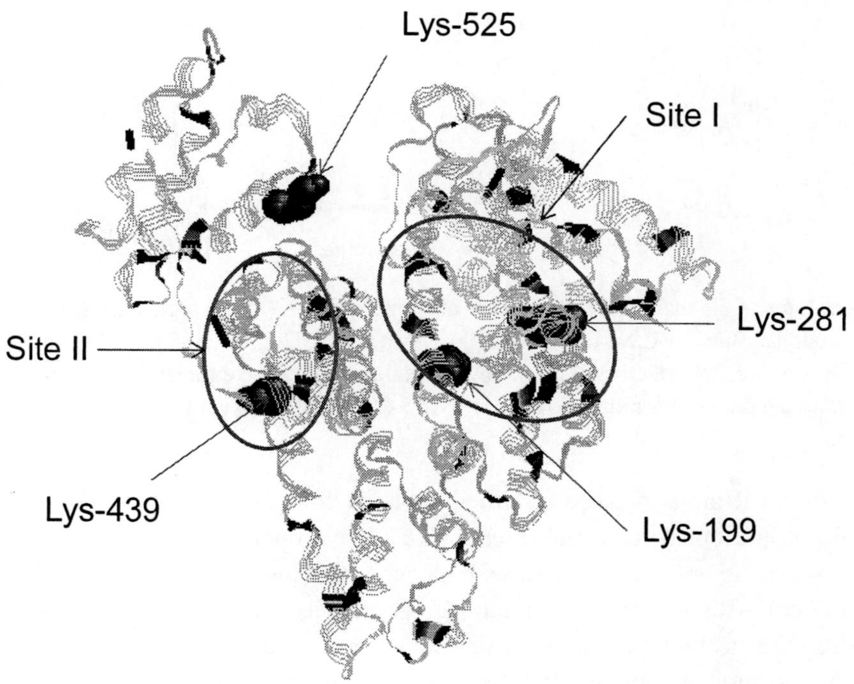

Figure 1. Human serum albumin with the two main drug binding sites and the location of the four lysine residues which are the main glycation sites. The structure of albumin is shown as strands using a PDB file (PDB ID: 2BXF).

Glycoxidation of HSA in both *in vitro* and *in vivo* models is associated with important structural modifications. In particular, the tertiary structure conformation, as probed by tryptophan fluorescence, is significantly affected by glycation. For HSA, the sole tryptophan, Trp-214, is located in domain II. Several studies have reported that tryptophan fluorescence is quenched in HSA that has been modified by glycation, reflecting a local unfolding around this residue [31, 32]. This structural modification has been shown to be accompanied by a partial opening of hydrophobic pockets for HSA that had been subjected to glycation *in vitro* for 21 days with glucose [31].

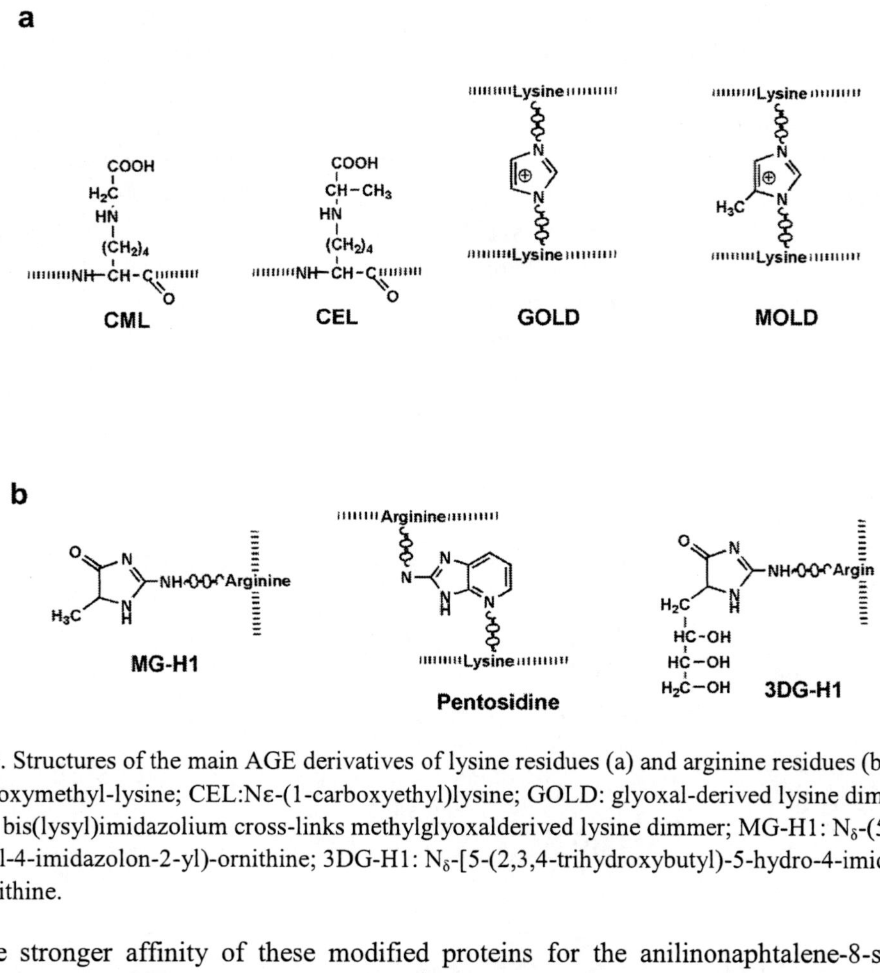

Figure 2. Structures of the main AGE derivatives of lysine residues (a) and arginine residues (b). CML: Nε-carboxymethyl-lysine; CEL:Nε-(1-carboxyethyl)lysine; GOLD: glyoxal-derived lysine dimmer; MOLD: bis(lysyl)imidazolium cross-links methylglyoxalderived lysine dimmer; MG-H1: N_δ-(5-hydro-5-methyl-4-imidazolon-2-yl)-ornithine; 3DG-H1: N_δ-[5-(2,3,4-trihydroxybutyl)-5-hydro-4-imidazolon-2-yl]ornithine.

The stronger affinity of these modified proteins for the anilinonaphtalene-8-sulfonate (ANS) probe is attributed to a higher exposure of hydrophobic sites to the solvent [33]. Conversely, when albumin is incubated with glucose for a longer period (more than 21 days), the penetration of ANS into the hydrophobic domains of the protein tend to be inhibited [34, 35]. The decrease in hydrophobicity in such long-term modified HSA is explained by the formation of the molten globule-like state and amyloid nanofibrils after 3 weeks and 20 weeks of incubation with glucose, respectively [33, 36]. In parallel to the quenching of tryptophan fluorescence, the formation of Amadori products and AGEs from HSA is associated with the formation of a glucose-derived fluorescence, called glycophores. The potential chemical nature of this product can be attributed to the presence of an aromatic nucleus or π conjugated systems, which could result in the development of fluorescence in some AGE products, such as GOLD or MOLD (Figure 2). Several fluorescence studies performed on glycated HSA have reported that glycophore formation is associated with an apparent reduction in tryptophan fluorescence during glycation [21].

2.2. Functional Analysis of Glycated HSA

Glycation has a determinant impact on the functional properties of HSA, which can be related to alterations in its conformation and/or net charge. These glycation-derived modifications that induce these conformational changes in the protein structure can also change the binding properties of HSA, although they sometimes occur at residues that are located some distance from the binding sites (Table 1). According to Shaklaï et al. (1984), glycation affects both the tertiary conformation of HSA and its binding functions [21]. For instance, in one study, the binding of tryptophan, a site II ligand, was found not to be affected by glycation in HSA [37], whereas, in another study, a lower affinity for this ligand was reported for HSA that had been glycated with low concentrations of glucose [38]. Similarly, the affinity of warfarin for the site I binding site has been shown to not be affected in the case of HSA glycated with glucose (2 mol glucose/mol albumin) [39], while a decrease in binding that was observed with early stage glycated HSA has been attributed to conformational changes or to steric hindrance of the protein in other studies [40, 41]. In contrast, the excessive glycation (60 days) of albumin with a large amount of glucose (9 mol glucose/mol albumin) has been shown to enhance the binding of warfarin to recombinant albumin [42]. Other ligands also have altered affinity for modified albumin when the protein is glycated. This is true for bilirubin, whose binding capacity has been shown to be reduced to about 50% in comparison with untreated albumin [21]. Similarly, monoacetyldiaminophenyl sulphone, a ligand, in which the location of its binding site is similar to that for bilirubin, also appears to present a decreased affinity [43].

Table 1. Effects of glycation on the binding properties of albumin

Compounds	Modification of binding affinity	Reference	Compounds	Modification of binding affinity	Reference
Site I			*Site II*		
Warfarin	→	[39]	Diazepam	→	[43]
	↑	[42]			
	↓	[40,41]	Ibuprofen	↓	[39, 41]
Furosemide	↓	[40]	Dansylsarcosine	↓	[44]
Dansylamide	→	[39]	Dansylproline	↓	[39]
Phenylbutazone	↓	[39, 40]	L-Tryptophan	↓	[38]
				→	[37]
Salicylate	↓	[40]	Flufenamic acid	↓	[39]
Bilirubin	↓	[32]	Naproxen	→	[40]
Other sites			*Aminoterminal*		
Captopril	→	[45]	Copper	↓	[46]
Lipoic acid				↑	[47]
Fatty acids	↓	[48]	Iron	↑	[47]

The binding of site II ligands, dansylsarcosine, dansylproline and ibuprofen (three major ligands) have been found to bind less strongly to modified albumin [39, 44]. In parallel, the binding of some ligands of Sudlow's site I or II ligands, including diazepam, naproxen and dansylamine has been shown to be similar in both modified and non-modified albumin [39, 40, 43]. Of the ligands with an affinity for albumin that are not affected by glycation, captopril appears to be an interesting therapeutic drug for preventing serious long-term diabetes-related complications. Indeed, this angiotensin-converting enzyme inhibitor has the ability to reduce the *in vitro* glycation of albumin and has similar binding affinities toward both native and glycated albumin. Through its unchanged pharmacokinetic parameters, captopril could provide essential protection for diabetic patients for maintaining the structural and functional properties of the albumin molecule [45]. Contradictory results are also observed for the binding of copper to aminoterminus binding sites. Argirova et al. suggested that early glycated albumin has a lower capacity for binding copper ions in strong complexes, with the consequence that the redox abilities of protein-copper complexes are reduced [46]. In contrast, in another study using BSA that had been extensively glycated with glucose (500 mM), an increase in the binding capacity for copper was found in comparison with native albumin, with the redox activity of copper bound to glycated albumin being maintained [47]. Further, a decrease in the binding affinities of glycated HSA toward several fatty acids was observed by a colorimetric determination of non-esterified fatty acids (NEFA), whereas with an ANS fluorimetric method no significant change was evidenced [48].

The clinical implication of albumin glycation is that the binding of HSA to some drugs may be altered in various stages of diabetes [49]. Included among the drugs whose affinity for albumin binding are impaired in diabetic patients include diazepam, sulfisoxazole, phenytoin, cyclosporine and valproic acid [50]. However these alterations are not always associated with an increase in the albumin glycation rate. For some of these drugs (sulfisoxazole, phenytoin), the reduction in drug binding is a directly consequence of albumin being altered by glycation [51, 52]. To the contrary, the decreased plasma binding for diazepam appears to be caused by enhanced free fatty acid levels, rather than by glycation [52, 53]. Similarly, the lower binding of valproic acid in diabetes patients involved glucose-independent modifications of the albumin molecule. The structural and functional alterations of glycated proteins, including albumin, have important clinical implications. Elevated levels of glycated albumin induce irreversible damage associated with the metabolic disorders observed in diabetes mellitus, such as retinopathy, nephropathy, neuropathy and coronary artery disease [54, 55].

The deleterious effects of glycated albumin and the resulting production of advanced glycation end-products have been highlighted in many *in vivo* and *in vitro* studies. For example, most of the available studies conducted by Cohen have shown the existence of a physiopathological connection between glycated albumin and diabetic renal deficiency and microangiopathy [56-58]. This involvement of Amadori products was confirmed by the fact that glycated albumin is preferentially transported across the renal glomerular capillaries and is taken up by mesanglial and epithelial cells, with a consequent increase in oxidative products that are strongly involved in nephropathy development [59]. The role of glycated albumin in diabetes-related vascular complications has also been established in the case of diabetic retinopathy [60]. In addition, there is increasing evidence that advanced glycation end products derived from albumin play a role in the promotion of cardiovascular diseases. Among its many properties, albumin also has anticoagulant and antithrombotic effects through its ability to limit platelet aggregation, with the consequence of preventing the progression of cardiovascular disease. Several *in vitro* studies have pointed out the implication of glycated albumin in the up-regulation of platelet activation and aggregation [61, 62]. Moreover, the hyper-aggregating effects of glycated albumin have also been demonstrated with erythrocytes [63]. In addition to its likely role in the pathogenesis of diabetic micro and macrovascular complications, the glycation of albumin can also affect glucose metabolism in both skeletal muscle and adipocyte cells [64-66]. For example, in mouse adipocyte cell lines, albumin-derived AGEs have been shown to trigger the generation of intracellular ROS leading to an inhibition of glucose uptake [67]. The attenuation of adipocyte insulin sensitivity in the presence of glycated proteins indicates their role in the development of obesity-related insulin resistance. The causal factor of insulin resistance is oxidative stress and intracellular ROS generation. Furthermore, it has been well established that advanced glycated proteins, including albumin, contribute to the oxidative modification of intracellular proteins in adipocyte cells [67, 68].

2.3. Therapeutic Applications of Glycated HSA

Numerous studies have reported that glycated albumin can be used as an alternative marker for glycemic control. As evidence for this, it has been shown that glycated albumin is strongly involved in the development of major diabetes complications, including nephropathy [69], retinopathy [70] and Alzheimer's disease [71]. In addition, in contrast to glycated hemoglobin (HbA1C), glycated albumin levels are associated with peripheral vascular calcification [72] and arterial stiffening [73], which are two of the most commonly observed diabetic complications. The results of recent clinical studies suggest that glycated albumin is an ideal marker of glycemic control in numerous physiopathological states, including hemodialysis patients [74], gastrectomized subjects [75], or for gestational diabetes [76]. All of these data support the utility of glycated albumin in the detection of short-term changes in glycemic control. However, in some pathological conditions that affect albumin metabolism, such as thyroid dysfunction, the nephrotic syndrome or liver cirrhosis, glycated albumin levels are not a suitable indicator for glucose excursion [77, 78]. Furthermore, the smoking status of diabetic patients has been identified as a significant negative explanatory variable, which affects the correlation between glycated albumin and plasma glucose levels [79]. Similarly, other studies have demonstrated that glycated albumin could be influenced by other physiological or pathological conditions, such as body mass index (BMI) [81] or the age of the diabetic patient [80].

3. Oxidation of HSA

Free radicals are a normal component of cellular oxygen metabolism in mammals. However, free radical associated damage is an important factor in many pathological processes. Posttranslational oxidative modifications of plasma proteins may serve as hallmarks of the severity of a disease and could result in alterations in protein function and structure. In fact, the oxidation of HSA is also closely related to the progress of several diseases. This section focuses on the characterization of albumin after oxidation by ROS in *in vitro* and *in vivo* studies and on the relevance of the products for several diseases.

3.1. Chemistry of Oxidized Amino Acid Residue

A typical example of a reversible modification is the mild oxidation of thiol groups, producing sulfenic acids and S-nitrosothiols, which represents a normal step in regulatory processes [82]. In contrast, Irreversible modification to sulfonic acid (SO_3H) causes loss of function and protein degradation and is generally considered to be a suicide function of target proteins that are relevant in antioxidant function in several diseases. Other major intermediates of protein oxidation involve the modification of methionine to methionine sulfoxide (MetO) or sulfone ($MetO_2$) or the oxidation of tyrosine by hypochlorous or hypobromous acids to give mono and dichloro-derivatives (Cl-Tyr, di-Cl-Tyr and Br-Tyr, di-Br-Tyr) or by peroxynitrite (NO_2-Tyr) [83]. Protein carbonylation is a less specific modification that involves a number of amino acid side chains, i.e., proline, arginine, lysine and threonine, oxidation to a ketone derivative. The presence of these stable final products along the protein chain is considered to serve as an indicator of the extent of oxidative damage [84-86]. In general, individual oxidants are known to generate different types of radicals. For example, all amino acid residues of proteins are potential targets for oxidation by $HO^{\bullet}$, which is generated upon exposure to ionizing radiation or by high concentrations of hydrogen peroxide and Cu^{2+} or Fe^{2+} [87-89]. However, these strong oxidation systems are generally associated with causing a stable loss of protein function, leading to the degradation of the modified protein.

A metal-catalyzed oxidation (MCO) system using very low concentrations of metal ions that result in mild oxidation was employed to evaluate the physiological significance of the impact of oxidation on proteins, which is a different process from the proteolytic attack that occurs in intracellular catabolism. Such an MCO system permits early alterations in the albumin molecule to be examined. Our findings indicate that multiple amino acid residues of HSA (Cys, Arg and Trp) are modified by the MCO system (Figure 3). These modifications have no detectable effect on the binding of drugs to site I in subdomain IIA. In contrast, the ligand-binding properties of the oxidized HSA for site II drugs and esterase-like activity are decreased, most probably due to conformational changes that occurred in subdomain IIIA [90]. In mice, the plasma half-life of oxidized HSA was substantially decreased and the clearance of oxidized HSA via uptake by the liver was markedly increased by 11-fold [91]. In addition, using a milder oxidation system, we were able to show that Cys and Met residues of HSA are modified by reaction with chloramines and hydrogen peroxide. These oxidations have no detectable effect on the ligand-binding properties, esterase-like activity, or the plasma half-life of HSA [90]. Another study showed that oxidation of key amino acid residues such as Trp, Arg, and Tyr may promote the elimination of HAS [92]. In fact, the oxidation of Arg-410 has been shown to promote the elimination of HSA from the circulatory system [93].

On the other hand, chloramines also react primarily with the free thiol of Cys-34 to form SO_3H, the stable adduct of the reaction. The reaction involving Cys-34 is first limited to the formation of a sulfenic (SOH) or a sulfinic acid (SO_2H) both being reversible intermediates that may shift back to the original mercapto-form or react with free glutathione or homocysteine to form non-mercapto albumin, which represents 5–15% of the total amount of the protein in serum [94].

The stable oxidation of the thiol group of Cys-34 to SO_3H results in the alteration in both the structural and functional characteristics of HSA making the protein to become more susceptible to trypsin digestion and an increase in the level of the non-oxidized counterpart, due to other major oxidation sites such as Arg, Trp and Tyr [84]. In fact, Kikugawa et al. proposed that, upon the oxidation of Trp, oxidized albumin becomes more susceptible to proteolytic cleavage by oxidized protein hydrolase [95].

Taking all of the above information into consideration, it appears that the preferred site of oxidation is dependent on the nature of the oxidant and the process relies on specific structural motifs in the protein to direct the oxidation (Figure 3) [96]. As shown in the in vitro oxidation model where oxidation changes HSA properties, the *in vivo* oxidation of HSA and its possible clinical significance should be evaluated.

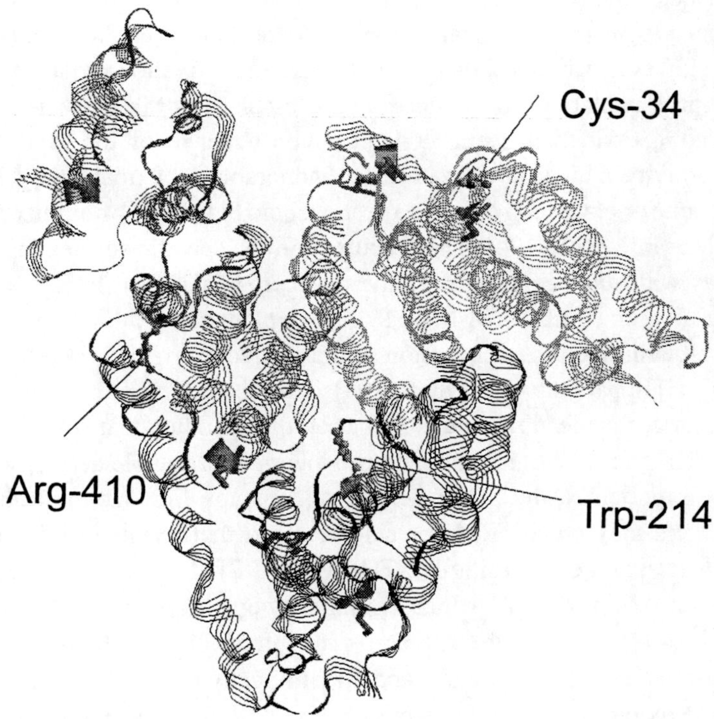

Figure 3. Possible major sites of oxidation of human serum albumin. The structure of albumin is shown as strands using PDB-file (1BMO). Cysteine-34 is particularly susceptible to oxidation. The six methionine residues are depicted in stick representations in blue.

3.2. Relationships between "*In Vitro*" and "*In Vivo*" Oxidized HSA

In recent studies, evidence suggesting that the oxidation of HSA also occurs *in vivo* has been reported. Previous reports, based on indirect evidence, indicated that only the partial oxidation of HSA occurs in cases of chronic renal failure (CRF) and diabetes mellitus [97-99]. Studies of patients with CRF are particularly relevant, since they typically have increased levels of human non-mercapt albumin (HNA) that may represent the product of the reaction of free plasma thiols, such as glutathione and/or homocysteine, with the sulfenic derivative of HSA. These studies indicated that HNA levels are correlated with the degree of renal dysfunction with a marked increase in patients who are undergoing hemodialysis (HD) [100]. We also showed that HSA is the major plasma protein target of oxidative stress in uremia, and that increased levels of carbonyl compounds are correlated with the HNA levels in uremic patients [101]. Therefore, there may be a variety of biological mechanisms responsible for these effects. In fact, we reported that oxidized HSA purified from HD patients showed, not only reduced structural and functional properties, but also acted as a mediator of the neutrophil activation state associated with chronic uremia [12]. On the other hand, other research groups have reported the presence of elevated levels of advanced oxidation products (AOPPs), leading to the formation of dityrosine and account for the excess oxidation of albumin, in several diseases including conditions involving or with already developed renal lesions such as diabetes mellitus and IgA glomerulonephritis. When present, AOPPs can be used to predict the progression of renal lesions, atherosclerotic cardiovascular events and death [102, 103]. However, structural data showing specific oxidation products of albumin in vivo are not currently available. The indication that albumin is the most important plasma component undergoing oxidation seems a logical consequence of the fact that it retains myeloperoxidase activity and is, for this reason, vulnerable to oxidation by hypochlorous acid. To date, only one attempt to arrive at a spectroscopic definition of oxidized albumin has been reported by Musante et al. [104]. They reported on the specific sulfonation of Cys-34 of albumin in patients with focal segmental glomerulosclerosis (FSGS). In fact, an analysis of plasma and serum albumin by LC-ESI-MS/MS first demonstrated the massive and stable in vivo oxidation of albumin with the conversion of the unique free –SH group of Cys-34 into a sulfonic group (SO_3H) in patients with FSGS [104]. This alteration produces relevant changes in the protein structure, thus leading to the formation of an adduct that is 48Da heavier than the native protein. Therefore, plasma oxidation produces a more stable derivative *in vivo* than *in vitro* and, most importantly, the SO_3H derivative does not dimerize but rather undergoes proteolysis. This event may be considered a suicide effect that results in HSA being cleared from the plasma. Kawakami et al., using LC-TOFMS and FTMS analysis [84], proposed that the increased level of oxidized HSA, which is primarily cysteinylated via the formation of a disulfide bond at Cys-34, results in the impaired function of HSA, including susceptibility to proteases, ligand-binding affinity and antioxidant activity. Collectively, these data demonstrate that the oxidation of Cys-34 occurs in vivo and, that this process is correlated, to some extent, with the progression of oxidative stress mediated diseases such as diabetes and renal failure. Thus, the redox states of Cys-34 of albumin might be useful biomarkers for the quantitative and qualitative evaluation of a diseased state.

3.3. Potential Therapeutic Applications of Oxidized HSA

Since oxidative stress causes molecular modifications of HSA, such as carbonylation, the formation of advanced glycation end products (AGEs) and AOPP, these stable final products along the protein chain are considered to be an indicator of the extent of oxidative damage [12, 85, 86]. However, since these markers are end products, sensitive markers that could be used to estimate the initial oxidative states continue to be needed. Oxidation of the amino acyl side chains of amino acids in proteins is an attractive biomarker for oxidative reactions because of the high specificity of the end products for specific oxidation pathways. In particular, the detection of oxidized cysteine residues currently constitutes the most sensitive and specific methods available for detecting end products of specific oxidative pathways [105].

Cys-34 is a particular redox sensitive site of albumin (Figure 3) and, therefore, it is reasonable to think of the redox state of albumin in terms of Cys-34. The chromatographic separation of albumin gives three fractions according to the state of Cys-34: (i) the free thiol form, referred to as HMA; (ii) the disulfide form with a small thiol compound, mainly cysteine or cysteinylglycine but also homocysteine and glutathione, referred to as human non-mercapt albumin1 (HNA-1); and (iii) a higher oxidized form in which cysteine has been oxidized to the level of a sulfinic or sulfonic acid (HNA-2) [106]. For example, in a healthy young person, HMA accounts for 70–80%, HNA-1 for 20–30%, and HNA-2 for about 2–5% of the total albumin present [107]. During the past 10 years, it has been shown that HMA and HNA are in equilibrium, depending on the surrounding redox state of Cys-34, and that their ratio varies depending on age and the diseased state. In fact, the HNA fraction has been found to increase during the course of several pathological conditions, such as renal dysfunction, various liver diseases, coronary artery disease, diabetes mellitus, senile cataracts, invasive surgery, and during intensive exercise and aging [108-112]. For example, the HNA fraction was negatively correlated with creatinine clearance in patients with renal dysfunction and, during intrauterine growth restriction, the fraction of HNA-2 in maternal plasma was about twice as high as that in a control plasma [101, 113].

In some cases, the treatment of a disorder becomes a source of oxidative stress, as in the intravenous administration of iron to HD patients, which was shown to increase both the carbonyl content of plasma proteins, especially albumin, and the HNA-1 and HNA-2 fractions in our studies [101]. We also reported that oxidative stress was significantly less evident, despite equivalent anti-anemic effects, when intravenous iron was given once weekly for 3 months, rather than three times weekly for 1 month. Based on these results, we reevaluated the conventional regimen for intravenous iron treatment in Japan from the viewpoint of oxidative stress (Figure 4) [114].

In general, variations in albumin fractions in patients and healthy subjects are quite high. Hence, it is difficult to draw conclusions from the momentary redox state of an albumin sample. In fact, while there is a highly significant correlation of the HNA-1 fraction with age, no such correlation was found for HNA-2 [108]. The variations are even higher for HNA-1. However, it should be noted that the quantification of the small HNA-2 peak is less accurate. To overcome these problems, we proposed that the oxidized albumin ratio (HNA/HMA), estimated by dividing the area under the peak for the reduced form (HMA) by the oxidized form (HNA-1+HNA-2), could be a useful index of oxidative stress. As shown in Figure 5, we have reevaluated the oxidized albumin ratio in several diseases and proposed that the oxidized albumin ratio might be a more useful tool for the systemic redox states in health, disease, and aging [107]. In Figure 5, we also demonstrated that AST-120 (Kremezin®), an oral carbonaceous adsorbent, that has been used in pre-dialysis, reduced the oxidized albumin ratio from 0.8-1.2 to 0.6-0.8 in CRF by removing uremic toxins from the systemic circulation [115]. We also found that Olmesartan, an inverse angiotensin II type 1 receptor antagonist, effectively lowers the extent of oxidized albumin ratio from 1.2-2.0 to 0.8-1.2 in HD patients [116]. Furthermore, we have proposed that the co-administration of low-molecular weight chitosans, a natural polymer derived from chitin, may inhibit neutrophil activation and the oxidation of HSA that is commonly observed in HD patients, resulting in a reduction of uremia-related oxidative stress [117-119]. In conclusion, these results suggest that the treatments where the levels of oxidized albumin ratio (HNA/HMA) are minimized may be beneficial in several diseases for preventing the onset and progression of serious complications, such as cardiovascular disease, which affects the prognosis for survival (Figure 6).

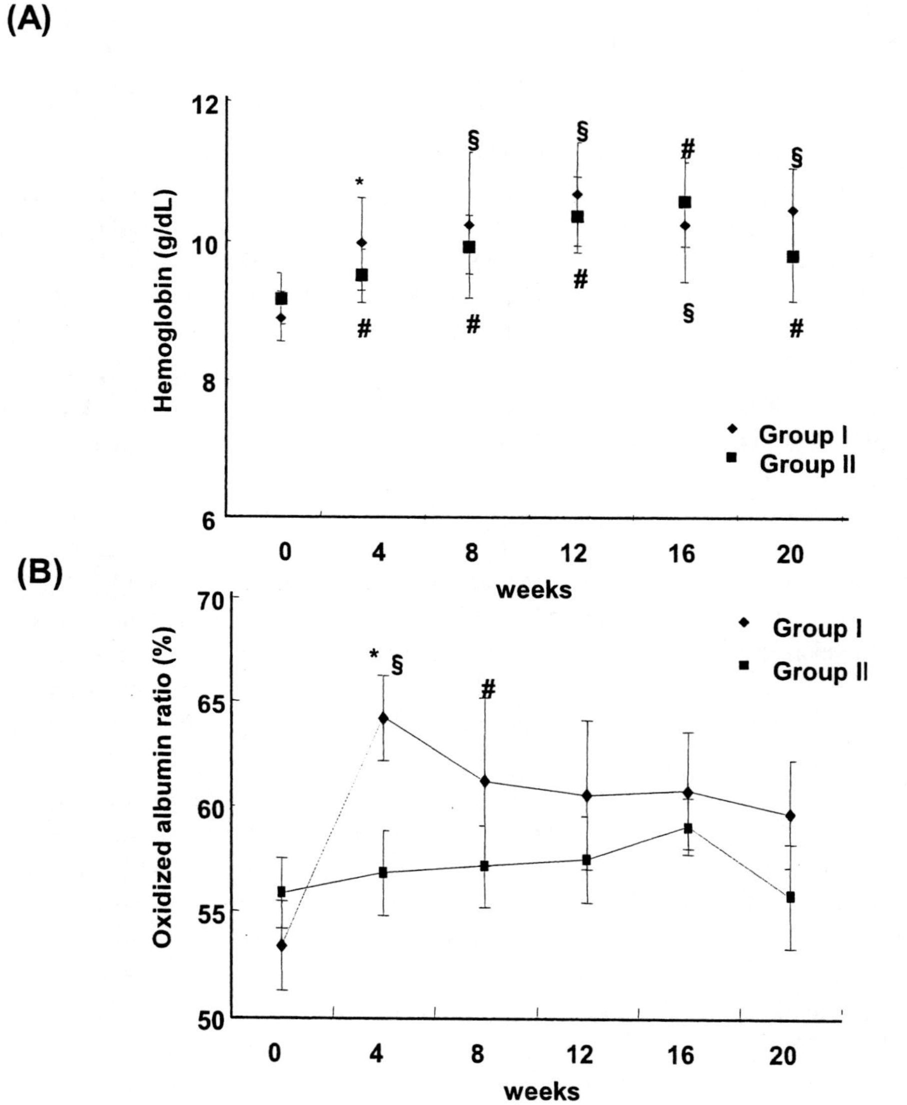

Figure 4. Effect of intravenous iron administration (IVIR) frequency on hemoglobin and oxidized HSA levels. Both the higher IVIR frequency group (group I: 40 mg of iron 3 times a week for 4 weeks, ♦) and the lower IVIR frequency group (group II: 40 mg of iron once a week for 12 weeks, ■) showed similar amelioration of hemoglobin levels (A) and oxidized HSA levels (B). * $P < 0.05$ as compared with 0 week (group II), # $P < 0.01$ as compared with 0 week (group I), and § $P < 0.01$ as compared with 0 week (group II). Values are expressed as mean ± SE; n =11 for each patient group.

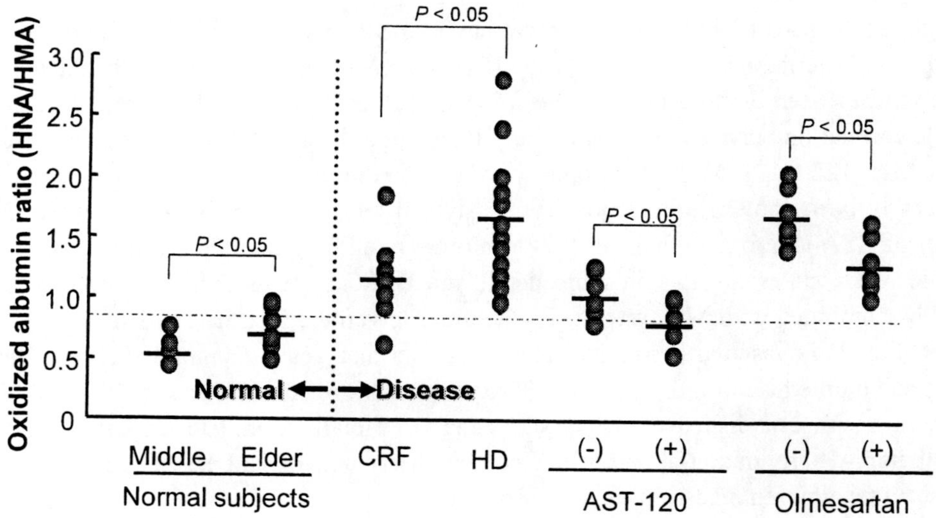

Figure 5. Changes in oxidized albumin ratios in healthy subjects and in different groups of patients. Values for each albumin fraction, HMA, HNA-1 and HNA-2, were estimated by dividing the area for each fraction by the total area for albumin. AST-120 (Kremezin®), an oral carbonaceous adsorbent, has been used in pre-dialysis, reduced oxidized albumin ratio in CRF.[112] Olmesartan (ARB), an inverse angiotensin II type 1 receptor antagonist, effectively lowers the extent of oxidized albumin ratio in HD patients. The information is taken from [98, 112, 113].

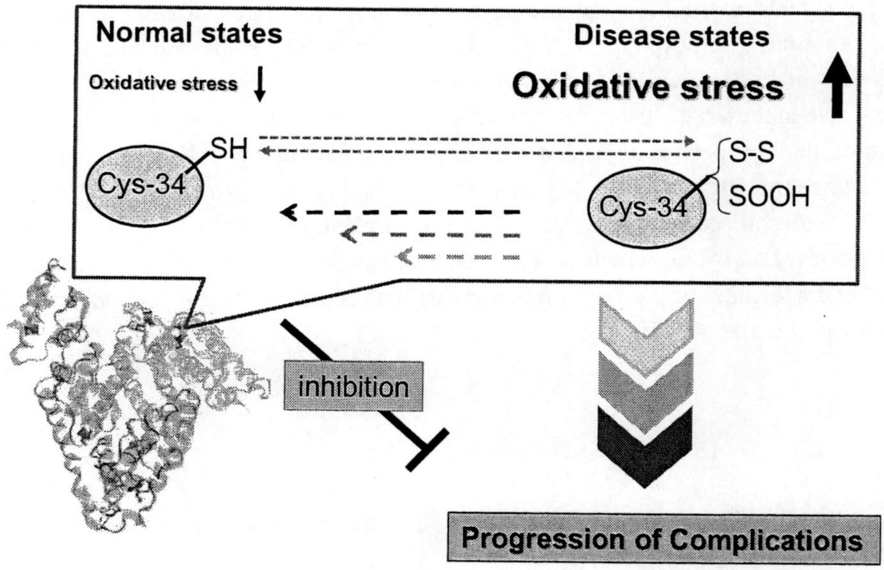

Figure 6. Proposed mechanism for the contribution of Cys-34 to the maintenance of homeostasis in the blood. Disease treatments where the level of Cys-34 oxidation is diminished may be beneficial for preventing the onset and progression of serious complications, which affects the prognosis for survival.

Recently, a new oxidative stress marker using the modification of HSA, ischemia modified albumin (IMA; Ischemia Technologies, Inc), measured by an albumin cobalt binding test (ACB; Ischemia Technologies), has been shown to be a marker of myocardial ischemia. ACB measures the ability of the amino terminus (N-terminus) of human albumin to bind exogenous cobalt. Under physiological conditions, transition metals can bind tightly to the exposed N-terminus of albumin [120, 121]. Given the fact that a significant negative correlation has been demonstrated between IMA and serum albumin levels in pregnancy, a possible correlation between HNA/HMA and IMA might lead to the more accurate maker of HNA/HMA [122, 123]. Mass spectrometry (MS) including LC-TOFMS and FTMS analysis has recently been shown to be a powerful analytical technique capable of the rapid online analysis of HSA isoforms in diabetes [124]. In order to analyze the information regarding the systemic redox states of HSA in more detail, the HPLC systems that are currently in use should be combined with MS systems. Furthermore, given the fact that oxidative stress, may be related to cardiovascular morbidity outcomes rather than actual albumin levels as reflected by hypoalbuminemia in HD patients [125], we believe that future studies involving populations with established oxidative stress may eventually stimulate the development of new albumin-determination-assays using combinations of MS and HPLC, and will allow more accurate albumin measurements to be obtained.

CONCLUSION

This review outlined the non-enzymatic glycation events that occur with proteins particularly for HSA, the major protein present in serum and is more sensitive to glycation reactions in a hyperglycaemic environment. We elaborated that structural and biological impact of the glycation of HSA alters the physiological properties of albumin, thus potentially altering its function. The new aspect of glycated albumin being employed as a future disease diagnostic marker better than HbA1c in diabetes associated complications is also discussed. This review also summarizes the redox properties of albumin, with a particular focus on Cys-34. The antioxidant properties of HSA are largely dependent on Cys-34 and its contribution to the maintenance of intravascular homeostasis, including the protection of the vascular endothelium under disease conditions related to oxidative stress. Further, the oxidized albumin ratio (HMA/HNA), which is a function of the redox states of Cys-34, might be a useful marker for evaluating systemic redox states that is useful for the evaluation of disease progression and therapeutic efficacy.

REFERENCES

[1] The Albumin Website. Available from: http://albumin.org (accessed February 12, 2015).

[2] Peters T Jr. All about albumin (1996). Academic Press, San Diego.

[3] Roche R, Rondeau P, Singh NR, Tarnus E, Bourdon E. The antioxidant properties of serum albumin (2008). *FEBS Lett.* 582, 1783-1787.

[4] Cha MK, Kim IH. Glutathione-linked thiol peroxidase activity of human serum albumin: a possible antioxidant role of serum albumin in blood plasma (1996). *Biochem. Biophys. Res. Commun.* 222, 619-625.

[5] Soriani M, Pietraforte D, Minetti M. Antioxidant potential of anaerobic human plasma: role of serum albumin and thiols as scavengers of carbon radicals (1994). *Arch. Biochem. Biophys*. 312, 180-188.

[6] Bourdon E, Blache D. The importance of proteins in defense against oxidation (2001). *Antioxid Redox Signal*. 3, 293-311.

[7] Brownlee M. Biochemistry and molecular cell biology of diabetic complications (2001). *Nature*. 414, 813-820.

[8] Furukawa S, Fujita T, Shimabukuro M, Iwaki M, Yamada Y, Nakajima Y, et al. Increased oxidative stress in obesity and its impact on metabolic syndrome (2004). *J. Clin. Invest*. 114, 1752-1761.

[9] Droge W. Free radicals in the physiological control of cell function (2002). *Physiol. Rev*. 82, 47-95.

[10] Vlassara H. Advanced glycation in health and disease: role of the modern environment (2005). *Ann. N Y Acad. Sci*. 1043, 452-460.

[11] Devangelio E, Santilli F, Formoso G, Ferroni P, Bucciarelli L, Michetti N, et al. Soluble RAGE in type 2 diabetes: Association with oxidative stress (2007). *Free Radic. Biol. Med*. 43, 511-518.

[12] Terawaki H, Yoshimura K, Hasegawa T, Matsuyama Y, Negawa T, Yamada K, et al. Oxidative stress is enhanced in correlation with renal dysfunction: examination with the redox state of albumin (2004). *Kidney Int*. 66, 1988-1993.

[13] Musante L, Bruschi M, Candiano G, Petretto A, Dimasi N, DelBoccio P, et al. Characterization of oxidation end product of plasma albumin 'in vivo' (2006). *Biochem. Biophys. Res. Commun*. 349, 668-673.

[14] Radi R, Beckman JS, Bush KM, Freeman BA. Peroxynitrite oxidation of sulfhydryls. The cytotoxic potential of superoxide and nitric oxide (1991). *J. Biol. Chem*. 266, 4244-4250.

[15] Finch JW, Crouch RK, Knapp DR, Schey KL. Mass spectrometric identification of modifications to human serum albumin treated with hydrogen peroxide (1993). *Arch. Biochem. Biophys*. 305, 595-599.

[16] Stamler JS, Simon DI, Osborne JA, Mullins ME, Jaraki O, Michel T, et al. S-nitrosylation of proteins with nitric oxide: synthesis and characterization of biologically active compounds (1992). *Proc. Natl. Acad. Sci. USA*. 89, 444-448.

[17] DeMaster EG, Quast BJ, Redfern B, Nagasawa HT. Reaction of nitric oxide with the free sulfhydryl group of human serum albumin yields a sulfenic acid and nitrous oxide (1995). *Biochemistry*. 34:11494-11499.

[18] Zhang H, Means GE. S-nitrosation of serum albumin: spectrophotometric determination of its nitrosation by simple S-nitrosothiols (1996). *Anal. Biochem*. 237, 141-144.

[19] Gatti RM, Radi R, Augusto O. Peroxynitrite-mediated oxidation of albumin to the protein-thiyl free radical (1994). *FEBS Lett*. 348, 287-290.

[20] Bourdon E, Loreau N, Lagrost L, Blache D. Differential effects of cysteine and methionine residues in the antioxidant activity of human serum albumin (2005). *Free Radic Res*. 39, 15-20.

[21] Shaklai N, Garlick RL, Bunn HF. Nonenzymatic glycosylation of human serum albumin alters its conformation and function (1984). *J. Biol. Chem*. 259, 3812-3817.

[22] Garlick RL, Mazer JS. The principal site of nonenzymatic glycosylation of human serum albumin in vivo (1983). *J. Biol. Chem*. 258, 6142-6146.

[23] Iberg N, Fluckiger R. Nonenzymatic glycosylation of albumin in vivo. Identification of multiple glycosylated sites (1986). *J. Biol. Chem.* 261, 13542-13545.

[24] Marashi SA, Safarian S, Moosavi-Movahedi AA. Why major nonenzymatic glycation sites of human serum albumin are preferred to other residues? (2005). *Med. Hypotheses.* 64, 881

[25] Kisugi R, Kouzuma T, Yamamoto T, Akizuki S, Miyamoto H, Someya Y, et al. Structural and glycation site changes of albumin in diabetic patient with very high glycated albumin (2007). *Clin. Chim. Acta.* 382, 59-64.

[26] Lapolla A, Fedele D, Reitano R, Arico NC, Seraglia R, Traldi P, et al. Enzymatic digestion and mass spectrometry in the study of advanced glycation end products/peptides (2004). *J. Am. Soc. Mass Spectrom.* 15, 496-509.

[27] Wa C, Cerny RL, Clarke WA, Hage DS. Characterization of glycation adducts on human serum albumin by matrix-assisted laser desorption/ionization time-of-flight mass spectrometry (2007). *Clin. Chim. Acta.* 385, 48-60.

[28] Ahmed N, Thornalley PJ. Peptide mapping of human serum albumin modified minimally by methylglyoxal in vitro and in vivo (2005). *Ann. NY Acad. Sci.* 1043, 260-266.

[29] Iijima K, Fujimoto D, Irie S. Amount of N(omega)-(Carboxymethyl)arginine generated in collagen and bovine serum albumin during glycation reactions is significantly different (2007). *Connect Tissue Res.* 48, 271-276.

[30] Mera K, Fujiwara Y, Otagiri M, Sakata N, Nagai R. Immunological detection of N omega-(Carboxymethyl)arginine by a specific antibody (2008). *Ann. NY Acad. Sci.* 1126, 155-157.

[31] Coussons PJ, Jacoby J, McKay A, Kelly SM, Price NC, Hunt JV. Glucose modification of human serum albumin: a structural study (1997). *Free Radic. Biol. Med.* 22, 1217-1227.

[32] Bourdon E, Loreau N, Blache D. Glucose and free radicals impair the antioxidant properties of serum albumin (1999). *FASEB J.* 13, 233-244.

[33] Sattarahmady N, Moosavi-Movahedi AA, Ahmad F, Hakimelahi GH, Habibi-Rezaei M, Saboury AA, et al. Formation of the molten globule-like state during prolonged glycation of human serum albumin (2007). *Biochim. Biophys. Acta.* 1770, 933-942.

[34] Chamani J, Moosavi-Movahedi AA, Rajabi O, Gharanfoli M, MomenHeravi M, Hakimelahi GH, et al. Cooperativealpha-helix formation of beta-lactoglobulin induced by sodium n-alkylsulfates (2006). *J. Colloid Interface Sci.* 293, 52-60.

[35] Zoellner H, Hou JY, Hochgrebe T, Poljak A, Duncan MW, Golding J, et al. Fluorometric and mass spectrometric analysis of nonenzymatic glycosylated albumin (2001). *Biochem. Biophys. Res. Commun.* 284, 83-89.

[36] Sattarahmady N, Moosavi-Movahedi AA, Habibi-Rezaei M, Ahmadian S, Saboury AA, H. Heli H, et al. Detergency effects of nanofibrillar amyloid formation on glycation of human serum albumin (2008). *Carbohydr. Res.* 343, 2229-2234.

[37] Bohney JP, Feldhoff RC. Effects of nonenzymatic glycosylation and fatty acids on tryptophan binding to human serum albumin (1992). *Biochem. Pharmacol.* 43, 1829-1834.

[38] Barzegar A, Moosavi-Movahedi AA, Sattarahmady N, Hosseinpour- Faizi MA, Aminbakhsh M, Ahmad F, et al. Spectroscopic studies of the effects of glycation of human serum albumin on L-Trp binding (2007). *Protein Pept Lett.* 14, 13-18.

[39] Okabe N, Hashizume N. Drug binding properties of glycosylated human serum albumin as measured by fluorescence and circular dichroism (1994). *Biol. Pharm Bull.* 17, 16-21.

[40] Koizumi K, Ikeda C, Ito M, Suzuki J, Kinoshita T, Yasukawa K, et al. Influence of glycosylation on the drug binding of human serum albumin (1998). *Biomed. Chromatogr.* 12, 203-210.

[41] Mera K, Takeo K, Izumi M, Maruyama T, Nagai R, Otagiri M. Effect of reactive-aldehydes on the modification and dysfunction of human serum albumin (2009). *J. Pharm. Sci.* 99, 1614-1625.

[42] Nakajou K, Watanabe H, Kragh-Hansen U, Maruyama T, Otagiri M. The effect of glycation on the structure, function and biological fate of human serum albumin as revealed by recombinant mutants (2003). *Biochim. Biophys. Acta.* 1623, 88-97.

[43] Karp WB, Kinsley M, Subramanyam SB, Robertson AF. Binding properties of glycosylated albumin and acetaldehyde albumin (1985). *Alcohol Clin. Exp. Res.* 9, 429-432.

[44] Worner W, Pfleiderer S, Kratzer W, Rietbrock N. Non-enzymatic glycation of human serum albumin: influence on the binding kinetics of the benzodiazepine binding sites (1990). *J. Clin. Chem Clin. Biochem.* 28, 527-531.

[45] Mariee AD, Al-Shabanah O. Protective ability and binding affinity of captopril towards serum albumin in an in vitro glycation model of diabetes mellitus (2006). *J. Pharm Biomed. Anal.* 41, 571-575.

[46] Argirova MD, Ortwerth BJ. Activation of protein-bound copper ions during early glycation: study on two proteins (2003). *Arch. Biochem. Biophys.* 420, 176-184.

[47] Eaton JW, Qian M. Interactions of copper with glycated proteins: possible involvement in the etiology of diabetic neuropathy (2002). *Mol. Cell Biochem.* 234, 135-142.

[48] Yamazaki E, Inagaki M, Kurita O, Inoue T. Kinetics of fatty acid binding ability of glycated human serum albumin (2005). *J. Biosci.* 30, 475-481.

[49] Barnaby OS, Cerny RL, Clarke W, Hage DS. Comparison of modification sites formed on human serum albumin at various stages of glycation (2011). *Clin. Chim. Acta.* 412, 277-285.

[50] MacKichan J. Influence of protein binding and use of unbound (free) drug concentrations. in: Burton M, Shaw L, Schentag J, Evans W. (Eds.) (2006). Applied Pharmacokinetics & Pharmacodynamics: Principles of Therapeutic Drug Monitoring. Lippincott Williams & Wilkins, 82-120.

[51] Kearns G, Kemp S, Turley C, Nelson D. Protein binding of phenytoin and lidocaine in pediatric patients with type I diabetes mellitus (1988). *Dev Pharmacol Ther.* 11, 14-23.

[52] Ruiz-Cabello F, Erill S. Abnormal serum protein binding of acidic drugs in diabetes mellitus (1984). *Clin. Pharmacol. Ther.* 36, 691-695.

[53] Erill S, Calva R. Post-translational changes of albumin as a cause of altered drug-plasma protein binding. in: M.M.E.S. (Ed.) (1986). Drug-Protein Binding. Praeger Publishers, New-York, 220-232.

[54] Cohen MP. Intervention strategies to prevent pathogenetic effects of glycated albumin (2003). *Arch. Biochem. Biophys.* 419, 25-30.

[55] Brownlee M. The pathological implications of protein glycation (1995). *Clin. Invest Med.* 18, 275-281.

[56] Chen S, Cohen MP, Ziyadeh FN. Amadori-glycated albumin in diabetic nephropathy: pathophysiologic connections (2000). *Kidney Int. Suppl.* 77, S40-S44.
[57] Cohen MP, Hud E, Wu VY. Amelioration of diabetic nephropathy by treatment with monoclonal antibodies against glycated albumin (1994). *Kidney Int.* 45, 1673-1679.
[58] Cohen MP, Masson N, Hud E, Ziyadeh F, Han DC, Clements RS. Inhibiting albumin glycation ameliorates diabetic nephropathy in the db/db mouse (2000). *Exp. Nephrol.* 8, 135-143.
[59] Ziyadeh FN, Cohen MP. Effects of glycated albumin on mesangial cells: evidence for a role in diabetic nephropathy (1993). *Mol. Cell Biochem.* 125, 19-25.
[60] Hasan NA. Effects of trace elements on albumin and lipoprotein glycation in diabetic retinopathy (2009). *Saudi Med. J.* 30, 1263-1271.
[61] Doweiko JP, Bistrian BR. The effect of glycosylated albumin on platelet aggregation (1994). *J. Parenter Enteral Nutr.* 18, 516-520.
[62] Rubenstein DA, Yin W. Glycated albumin modulates platelet susceptibility to flow induced activation and aggregation (2009). *Platelets.* 20, 206-215.
[63] Candiloros H, Muller S, Ziegler O, Donner M, Drouin P. Role of albumin glycation on the erythrocyte aggregation: an in vitro study (1996). *Diabet Med.* 13, 646-650.
[64] Miele C, Riboulet A, Maitan MA, Oriente F, Romano C, Formisano P, et al. Human glycated albumin affects glucose metabolism in L6 skeletal muscle cells by impairing insulin-induced insulin receptor substrate (IRS) signaling through a protein kinase C alphamediated mechanism (2003). *J. Biol. Chem.* 278, 47376-47387.
[65] Unoki H, Bujo H, Yamagishi S, Takeuchi M, Imaizumi T, Saito Y. Advanced glycation end products attenuate cellular insulin sensitivity by increasing the generation of intracellular reactive oxygen species in adipocytes (2007). *Diabet Res. Clin. Pract.* 76, 236-244.
[66] Unoki H, Yamagishi S. Advanced glycation end products and insulin resistance (2008). *Curr. Pharm Des.* 14, 987-989.
[67] Singh NR, Rondeau P, Bourdon E. Identification of up-regulated low molecular weight proteins in human adipocytes treated with glycoxidized albumin (2010). *Open Obes J.* 2, 110-115.
[68] Singh NR, Rondeau P, Hoareau L, Bourdon E. Identification of preferential protein targets for carbonylation in human mature adipocytes treated with native or glycated albumin (2007). *Free Radic. Res.* 41, 1078-1088.
[69] Gugliucci A, Glycation as the glucose link to diabetic complications (2000). *J. Am. Osteopath Assoc.* 100, 621-634.
[70] Okumura A, Mitamura Y, Namekata K, Nakamura K, Harada C, Harada T. Glycated albumin induces activation of activator protein-1 in retinal glial cells (2007). *Jpn J. Ophthalmol.* 51, 236-237.
[71] Shuvaev VV, Laffont I, Serot JM, Fujii J, Taniguchi N, Siest G. Increased protein glycation in cerebrospinal fluid of Alzheimer's disease (2001). *Neurobiol. Aging.* 22 , 397-402.
[72] Yamada S, Inaba M, Shidara K, Okada S, Emoto M, Ishimura E, et al. Association of glycated albumin, but not glycated hemoglobin, with peripheral vascular calcification in hemodialysis patients with type 2 diabetes (2008). *Life Sci.* 83, 516-519.
[73] Kumeda K, Inaba M, Shoji S, Ishimura E, Inariba H, Yabe S, et al. Significant correlation of glycated albumin, but not glycated haemoglobin, with arterial stiffening

in haemodialysis patients with type 2 diabetes (2008). *Clin. Endocrinol.* (Oxf.) 69, 556-561.

[74] Nagayama H, Inaba M, Okabe R, Emoto M, Ishimura E, Okazaki S, et al. Glycated albumin as an improved indicator of glycemic control in hemodialysis patients with type 2 diabetes based on fasting plasma glucose and oral glucose tolerance test (2009). *Biomed. Pharmacother.* 63,236-240.

[75] Koga K, Murai J, Saito H, Mukai M, Matsumoto S, Kasayama S, et al. Glycated albumin levels are higher relative to glycated haemoglobin levels in gastrectomized subjects (2009). *Ann. Clin. Biochem.* 47, 39-43.

[76] Hashimoto K, Osugi T, Noguchi S, Morimoto Y, Wasada K, Imai S, et al. A1C but not serum glycated albumin is elevated because of iron deficiency in late pregnancy in diabetic women (2008). *Diabet Care.* 33, 509-511.

[77] Koga M, Kasayama S, Kanehara H, Bando Y. CLD (chronic liver diseases)- HbA1C as a suitable indicator for estimation of mean plasma glucose in patients with chronic liver diseases (2008). *Diabet Res. Clin. Pract.* 81, 258-262.

[78] Koga M, Murai J, Saito H, Matsumoto S, Kasayama S. Effects of thyroid hormone on serum glycated albumin levels: study on non-diabetic subjects (2009). *Diabet Res. Clin. Pract.* 84, 163-167.

[79] Koga M, Saito H, Mukai M, Otsuki M, Kasayama S. Serum glycated albumin levels are influenced by smoking status, independent of plasma glucose levels (2009). *Acta Diabetol.* 46, 141-144.

[80] Koga M, Matsumoto S, Saito H, Kasayama S. Body mass index negatively influences glycated albumin, but not glycated hemoglobin, in diabetic patients (2006). *Endocr. J.* 53, 387-391.

[81] Nishimura R, Kanda A, Sano H, Matsudaira T, Miyashita Y, Morimoto A, et al. Glycated albumin is low in obese, non-diabetic children (2006). *Diabet Res. Clin. Pract.* 71, 334-338.

[82] Carballal S, Radi R, Kirk MC, Barnes S, Freeman BA, Alvarez B. Sulfenic acid formation in human serum albumin by hydrogen peroxide and peroxynitrite (2003). *Biochemistry.* 42, 9906–9914.

[83] Dalle-Donne I, Scaloni A, Giustarini D, Cavarra E, Tell G, Lungarella G, et al. Proteins as biomarkers of oxidative/nitrosative stress in diseases: the contribution of redox proteomics (2005). *Mass Spectrom Rev.* 24, 55-99.

[84] Kawakami A, Kubota K, Yamada N, Tagami U, Takehana K, Sonaka I, et al. Identification and characterization of oxidized human serum albumin. A slight structural change impairs its ligand-binding and antioxidant functions (2006). *FEBS J.* 273, 3346-3357.

[85] Carr AC, Hawkins CL, Thomas SR, Stocker R, Frei B. Relative reactivities of N-chloramines and hypochlorous acid with human plasma constituents (2001). *Free Radic. Biol. Med.* 30, 526-536.

[86] Salavej P, Spalteholz H, Arnhold J. Modification of amino acid residues in human serum albumin by myeloperoxidase (2006). *Free Radic. Biol. Med.* 40, 516-525.

[87] Garrison WM. Reaction mechanisms in the radiolysis of peptides, polypeptides, and proteins (1987). *Chem. Rev.* 87, 381-398.

[88] Schuessler H, Schilling K. Oxygen effect in radiolysis of proteins. Part 2.Bovine serum albumin (1984). *Int. J. Radiat. Biol.* 45, 267-281.

[89] Neuzil J, Gebicki JM, Stocker R. Radical-induced chain oxidation of proteins and its inhibition by chain-breaking antioxidants (1993). *Biochem J.* 293, 601-606.

[90] Anraku M, Yamasaki K, Maruyama T, Kragh-Hansen U, Otagiri M. Effect of oxidative stress on the structure and function of human serum albumin (2001). *Pharm. Res.* 18, 632-639 (2001).

[91] Iwao Y, Anraku M, Hiraike M, Kawai K, Nakajou K, Kai T, et al. Suenaga, A. The structural and pharmacokinetic properties of oxidized human serum albumin, advanced oxidation protein products (AOPP) (2006). *Drug Metab. Pharmacokinet.* 21, 140-146.

[92] Anraku M, Kragh-Hansen U, Kawai K, Maruyama T, Yamasaki Y, Takakura Y, et al. Validation of the chloramine-T induced oxidation of human serum albumin as a model for oxidative damage in vivo (2003). *Pharm. Res.* 20, 684-692.

[93] Iwao Y, Anraku M, Yamasaki K, Kragh-Hansen U, Kawai K, Maruyama T, et al. Oxidation of Arg-410 promotes the elimination of human serum albumin (2006). *Biochim. Biophys. Acta.* 1764, 743-749.

[94] Meucci E, Mordente A, Martorana GE. Metal-catalyzed oxidation of human serum albumin: conformational and functional changes. Implications in protein aging (1991). *J. Biol. Chem.* 266, 4692-4699.

[95] Fujino T, Kojima M, Beppu M, Kikugawa K, Yasuda H, Takahashi K. Identification of the cleavage sites of oxidized protein that are susceptible to oxidized protein hydrolase (OPH) in the primary and tertiary structures of the protein (2000). *J. Biochem.* 127, 1087-1093.

[96] Temple A, Yen TY, Gronert S. Identification of specific protein carbonylation sites in model oxidations of human serum albumin (2006). *J. Am. Soc. Mass Spectrom.* 17, 1172-1180.

[97] Himmelfarb J, McMonagle E. Albumin is the major plasma protein target of oxidant stress in uremia (2001). *Kidney Int.* 60, 358-363.

[98] Suzuki E, Yasuda K, Takeda N, Sakata S, Era S, Kuwata K, et al. Increased oxidized form of human serum albumin in patients with diabetes mellitus (1992). *Diabetes Res. Clin. Pract.* 18, 153-158.

[99] Soejima A, Matsuzawa N, Hayashi T, Kimura R, Ootsuka T, Fukuoka K, Yamada A, et al. Alteration of redox state of human serum albumin before and after hemodialysis (2004). *Blood Purif.* 22, 525-529.

[100] Terawaki H, Yoshimura K, Hasegawa T, Matsuyama Y, Negawa T, Yamada K, et al. Oxidative stress is enhanced in correlation with renal dysfunction: examination with the redox state of albumin (2004). *Kidney Int.* 66, 1988-1993.

[101] Anraku M, Kitamura K, Shinohara A, Adachi M, Suenaga A, Maruyama T, et al. Intravenous iron administration induces oxidation of serum albumin in hemodialysis patients (2004). *Kidney Int.* 66, 841-848.

[102] Descamps-Latscha B, Witko-Sarsat V, Nguyen-Khoa T, Nguyen AT, Gausson V, Mothu N, et al. Advanced oxidation protein products as risk factors for atherosclerotic cardiovascular events in nondiabetic predialysis patients (2005). *Am. J. Kidney Dis.* 45, 39-47.

[103] Descamps-Latscha B, Witko-Sarsat V, Nguyen-Khoa T, Nguyen AT, Gausson V, Mothu N, et al. Early prediction of IgA nephropathy progression: Proteinuria and AOPP are strong prognostic markers (2004). *Kidney Int.* 66, 1606-1612.

[104] Musante L, Candiano G, Petretto A, Bruschi M, Dimasi N, Caridi G, et al. Active focal segmental glomerulosclerosis is associated with massive oxidation of plasma albumin (2007). *J. Am. Soc. Nephrol.* 18, 799-810.
[105] Heinecke JW, Hsu FF, Crowley JR, Hazen SL, Leeuwenburgh C, Mueller DM, et al. Detecting oxidative modification of biomolecules with isotope dilution mass spectrometry: sensitive and quantitative assays for oxidized amino acids in proteins and tissues (1999). *Methods Enzymol.* 300, 124-144.
[106] Soejima A, Kaneda F, Manno S, Matsuzawa N, Kouji H, Nagasawa T, et al. Useful markers for detecting decreased serum antioxidant activity in hemodialysis patients (2002). *Am. J. Kidney Dis.* 39, 1040-1046.
[107] Oettl K, Marsche G. Redox state of human serum albumin in terms of cysteine-34 in health and disease (2010). *Methods Enzymol.* 474, 181-195.
[108] Sogami M, Nagoka S, Era S, Honda M, Noguchi K. Resolution of human mercapt- and nonmercaptalbumin by high-performance liquid chromatography (1984). *Int. J. Pept. Protein Res.* 24, 96-103.
[109] Hayakawa A, Kuwata K, Era S, Sogami M, Shimonaka H, Yamamoto M, et al. Alteration of redox state of human serum albumin in patients under anesthesia and invasive surgery (1997). *J. Chromatogr. B. Biomed. Sci. Appl.* 698, 27-33.
[110] Era S, Kuwata K, Imai H, Nakamura K, Hayashi T, Sogami M. Age-related change in redox state of human serum albumin (1995). *Biochim. Biophys. Acta.* 1247, 12-16.
[111] Kadota K, Yui Y, Hattori R, Murohara Y, Kawai C. Decreased sulfhydryl groups of serum albumin in coronary artery disease (1991). *Jpn Circ. J.* 55, 937-941.
[112] Imai H, Hayashi T, Negawa T, Nakamura K, Tomida M, Koda K, et al. Strenous exercise-induced change in redox state of human serum albumin during intensive kendo training (2002). *Jpn J. Physiol.* 52, 135-140.
[113] Bar-Or D, Heyborne KD, Bar-Or R, Rael LT, Winkler JV, Navot D. Cysteinylation of maternal plasma albumin and its association with intrauterine growth restriction (2005). *Prenat Diagn.* 25, 245-249.
[114] Anraku M, Kitamura K, Shintomo R, Takeuchi K, Ikeda H, Nagano J, et al. Effect of intravenous iron administration frequency on AOPP and inflammatory biomarkers in chronic hemodialysis patients: a pilot study (2008). *Clin. Biochem.* 41, 1168-1174.
[115] Shimoishi K, Anraku M, Kitamura K, Tasaki Y, Taguchi K, Hashimoto M, et al. An oral adsorbent, AST-120 protects against the progression of oxidative stress by reducing the accumulation of indoxyl sulfate in the systemic circulation in renal failure (2007). *Pharm. Res.* 24, 1283-1289.
[116] Kadowaki D, Anraku M, Tasaki Y, Kitamura K, Wakamatsu S, Tomita K, et al. Effect of olmesartan on oxidative stress in hemodialysis patients (2007). *Hypertens Res.* 30, 395-402.
[117] Anraku M, Kabashima M, Namura H, Maruyama T, Otagiri M, Gebicki JM, et al. Antioxidant protection of human serum albumin by chitosan (2008). *Int. J. Biol. Macromol.* 43, 159-164.
[118] Anraku M, Fujii T, Furutani N, Kadowaki D, Maruyama T, et al. Antioxidant effects of a dietary supplement: reduction of indices of oxidative stress in normal subjects by water-soluble chitosan (2009). *Food Chem. Toxicol.* 47, 104-109.

[119] Anraku M, Fujii T, Kondo Y, Kojima E, Hata T, Tabuchi N, et al. Antioxidant properties of high molecular weight dietary chitosan in vitro and in vivo (2011). *Carbohydr. Poly.* 83, 501-505.

[120] Bar-Or D, Lau E, Winkler JV. A novel assay for cobalt-albumin binding and its potential as a marker for myocardial ischemia: a preliminary report (2000). *J. Emerg. Med.* 19, 311-315.

[121] Sadler PJ, Tucker A, Viles JH. Involvement of a lysine residue in the N-terminal Ni_2 and Cu_2 binding site of serum albumins: comparison with Co_2, Cd_2, Al_3. *Eur. J. Biochem.* 220, 193-200.

[122] Roy D, Quiles J, Gaze DC, Collinson P, Kaski JC, Baxter GF. Role of reactive oxygen species on the formation of the novel diagnostic marker ischaemia modified albumin (2006). *Heart.* 92, 113-114.

[123] Guven S, Alver A, Mentese A, Ilhan FC, Calapoglu M, Unsal MA. The novel ischemia marker 'ischemia-modified albumin' is increased in normal pregnancies (2009). *Acta Obstet. Gynecol. Scand.* 88, 479-482.

[124] Boisvert MR, Koski KG, Skinner CD. Increased oxidative modifications of amniotic fluid albumin in pregnancies associated with gestational diabetes mellitus (2010). *Anal. Chem.* 82, 1133-1137.

[125] Michelis R, Kristal B, Snitkovsky T, Sela S. Oxidative modifications impair albumin quantification (2010). *Biochem. Biophys. Res. Commun.* 401, 137-142.

In: Human Serum Albumin (HSA)
Editor: Travis Stokes

ISBN: 978-1-63482-963-2

Chapter 6

ANALYSIS OF THE ORGANIC SOLVENT EFFECT ON THE HYDRATION AND STRUCTURE OF HUMAN SERUM ALBUMIN BY INFRARED SPECTROSCOPY

Vladimir A. Sirotkin*
Kazan Federal University, A. M. Butlerov Institute of Chemistry,
Kazan, Russia

ABSTRACT

The effect of organic solvent on the hydration and structure of human serum albumin was characterized using infrared spectroscopy in the thermodynamic water activity range from 0 to 0.98 at 25°C. Dioxane was used as a model organic solvent. This organic solvent may be considered as an informative molecular probe for analyzing the effect of hydrogen bonding and hydrophobic interactions on the hydration and structure of proteins.

The obtained results show that the hydration and structure of human serum albumin depend markedly on how the protein has been hydrated – whether in the presence or in the absence of organic solvent. Two organic solvent effects on water sorption by human serum albumin were observed:

1. At low water activity (a_w<0.5), the water adsorption branch obtained in the presence of organic solvent is similar to that measured in the absence of dioxane.
2. At high water activity (a_w>0.5), the water sorption isotherm in the presence of dioxane lies above the corresponding isotherm for pure water. This result corresponds to the organic solvent – assisted effect on water binding by human serum albumin.

Changes in the structure of human serum albumin were determined from infrared spectra by analyzing the structure of amide I band. It was found that, at low water activity, the protein-protein contacts in the dried protein largely govern its thermodynamic and structural properties. At high water activity, the protein state is

* Kazan Federal University, A.M. Butlerov Institute of Chemistry, Kremlevskaya str., 18, Kazan, 420008, Russia; E-mail: vsir@mail.ru.

determined by the protein-organic solvent and protein-water interactions. The results from the thermodynamic and structural measurements were analysed to give a unified picture of the hydration process in the absence and presence of organic solvent.

1. Introduction

The hydration of proteins is a phenomenon of considerable fundamental importance and practical interest. It is well known that the interaction of water with proteins plays a key role in determining the structure, stability and functions of proteins [1-2]. Knowledge of processes occurring upon hydration of protein macromolecules is also very important in biotechnological and pharmaceutical applications of proteins such as their use as biocatalysts, biosensors and selective adsorbents in low-water media, including organic solvents, ionic liquids and supercritical fluids [3-8]. For example, there are many advantages in employing organic solvents for biocatalysis, including high solubility of reagents, synthesis of useful chemicals (chiral drug molecules, emulsifiers, modified fats, flavour esters), suppression of undesirable side reactions caused by water and increased thermostability. However, in general, the protein activity in organic solvents is a complex function of the water content in the organic media. A number of works [6-8] show that the hydration "history" is an important factor controlling the properties of proteins in organic media. This implies that kinetic limitations play an important role in the behaviour of proteins in water-restricted systems.

Consequently, studying the interactions of proteins with water in the presence of organic media may facilitate the development of novel efficient biocatalytic/biotechnological systems. In addition, such fundamental studies make it possible to extend our knowledge about protein macromolecules and the role that water plays in biocatalysis/protein stability, thereby providing information that is difficult or even impossible to obtain from classical experiments in aqueous media.

Hence, there is a considerable interest in the development of the experimental approaches by which both the thermodynamic and structural properties of the hydration of proteins in the presence of some low molecular additives, including organic solvents, may be obtained simultaneously.

The measurement of water sorption isotherms is one of the effective experimental approaches to obtain thermodynamic characteristics (water sorption equilibrium constant, water adsorption enthalpy, monolayer capacity) of the protein-water interactions in various environments [1, 2]. The hydration of proteins in organic solvents was investigated in many studies [8-14]. It was concluded that when the water sorption is plotted against the thermodynamic water activity (a_w), the sorption isotherms obtained in organic solvents are similar to those recorded in the absence of organic liquids at low a_w values [12, 14]. At higher water activities (a_w>0.4), the organic solvents caused deviations. Hydrophobic solvents (for example, benzene) usually increased the amount of water bound to the protein (at fixed a_w) while hydrophilic solvents (for example, ethanol and propanol-1) had the opposite effect.

Infrared spectroscopy is one of the effective methods for analyzing the structure of proteins in various environments, including aqueous and non-aqueous media [15-21]. This method has been successfully used in studying the secondary structure of proteins in various states, including protein films with various degrees of humidity and solid protein preparations immersed into pure organic solvents and water - organic mixtures. Infrared spectroscopy is also effective in studying the hydration of proteins.

In this study the effect of organic solvent on the protein hydration was investigated by FTIR spectroscopy in the water activity range from 0 to 0.98 at 25°C. The sorption data were compared with the corresponding structural changes that occur on the interaction of the protein with water and/or organic molecules. The aim of this FTIR spectroscopic study is to elucidate the effect of organic solvent molecules on the hydration and structure of the protein mcaromolecule.

Human serum albumin (HSA) is the most abundant in blood serum and plays a number of important biological roles, including the divalent cation transport, fatty acid and drug complexation and transport [22]. It consists of 585 amino acid residues. HSA is an example of a predominantly α-helix protein.

The choice of dioxane as a model was determined by the following reasons:

A) Dioxane is a water-miscible organic solvent. Therefore, it is possible to study the effects of this low molecular organic substance on the hydration and structure of proteins in the entire range of thermodynamic water activity.
B) It is capable of forming strong hydrogen bonds with various hydrogen bond donors. However, in contrast to water, it has no evident hydrogen bond donating ability.
C) Each dioxane molecule contains four hydrophobic CH_2-groups. Therefore, it is expected that there are considerable differences between the conformational changes and hydration-dehydration of proteins studied in the presence and absence of dioxane due to the hydrophobic interactions.

This means that this model organic solvent may be considered as an effective molecular probe for analyzing the effect of hydrogen bonding and hydrophobic interactions on the hydration and structure of proteins.

2. Experimental

2.1. Materials

Human serum albumin (Sigma, Product No. A 1887, essentially fatty acid free, protein content more than 96-99%) was used without further purification.

2.2. Solvents

1,4-Dioxane (reagent grade, purity >99%) was purified and dried according to the recommendations [23]. Water used was doubly distilled.

2.3. Thermodynamic Activity of Water in Organic Solvent

It is well known that water – organic mixtures are typical examples of nonideal systems where the thermodynamic characteristics of the state of water differ for the same molar ratio of components due to strong specific interactions between components (first of all, hydrogen bonding). To apply a correction for nonideality of a system, the concentrations of the substances are recalculated into thermodynamic activity scale. The thermodynamic activity of water (or organic solvent) is related to the chemical potential of the solute as (Eq. 1):

$$\mu_i = \mu_i^o + RT \ln a_i \qquad (1)$$

where μ_i^o is the standard chemical potential, and a_i is the activity of the *i*th component in the solution. Accordingly, the Gibbs energy change (ΔG) for the transfer of one mole of water from pure liquid state to aqueous organic medium with a mole fraction of water x_w equals (Eq. 2):

$$\Delta G = RT \ln a_W \qquad (2)$$

Therefore, the use of the thermodynamic activity scale allows one to normalize the concentration dependencies of characteristics of the effects under study with the respect to the Gibbs energy of water transfer from pure liquid to an water – organic medium.

Water activity (a_W) in organic solvent was calculated using Eq. 3:

$$a_W = \gamma_W x_W \qquad (3)$$

where x_W is the mole fraction of water in the solution and γ_W is the activity coefficient of water (in mole fractions; the standard state is pure water).

Water activity coefficients (γ_W) were calculated using literature data [24] on the vapor-liquid equilibrium according to Eq. 4.

$$\gamma_w = \frac{y_w P_{tot}}{x_w P_w^o} \qquad (4)$$

where y_W is the measured mole fraction of water in vapor phase, P_t is the total pressure, P_o is the saturated vapor pressure of pure water at the same temperature and x_W is the mole fraction of water in the liquid phase.

Figure 1 shows the thermodynamic activity of water plotted versus the water mole fraction in organic solvent at 25°C.

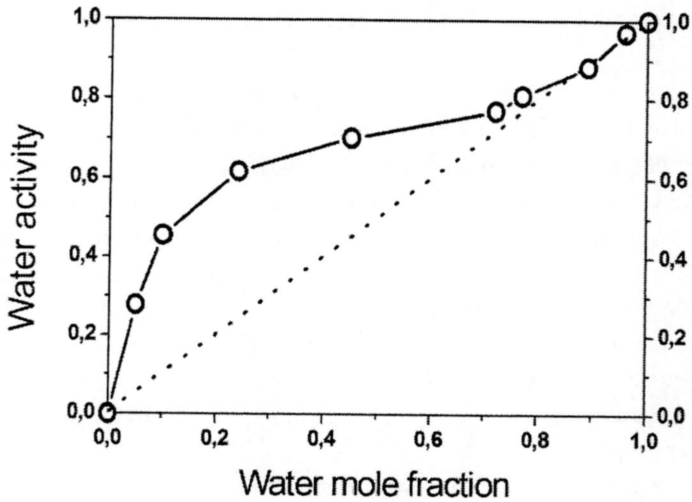

Figure 1. Thermodynamic activity of water plotted versus the water mole fraction in organic solvent at 25°C. The dashed lines correspond to the ideal binary mixture. Modified data from [24].

2.4. Spectroscopic Measurements

The FTIR spectra were recorded with a MAGNA 550 spectrometer from Nicolet. The infrared spectra were measured at 25°C as described previously [20, 21]. Each spectrum was obtained by coadding 256 scans at a spectral resolution of 2 cm^{-1}. The sample chamber was purged with dry and carbon dioxide free air. If required, the spectra were corrected for water vapour contribution to obtain a flat baseline between 2200 and 1850 cm^{-1}.

The infrared spectra were obtained with glassy like protein films casted from 2% (w/v) water solution onto the CaF_2 window at room humidity and temperature. Schematic representation of the experimental set-up for FTIR spectroscopic measurements of the structural and sorption characteristics of proteins is given in [21]. After mounting the window in the sample cell, the film was dehydrated by flushing air dried over P_2O_5 powder. Relative water vapour pressure over P_2O_5 at 25°C does not exceed 0.01 [25]. The protein films were flushed until no further spectral changes were detected in the 3450 cm^{-1} water absorbance region and amide A contour on this side represented a smooth line without any visible shoulders. The spectrum of this sample was used as a reference spectrum for calculation of the difference spectra.

The initially dehydrated protein film was exposed to pure water vapor or water-organic vapor mixtures. In the first case, pure water vapor consecutively flowed through the thermostated glass tube with drying agent (P_2O_5), saturator filled with pure liquid water and then through the measuring cell containing the protein sample. The temperature of sample cell was 25°C. The water activity (a_w) in vapor phase was adjusted by changing the difference between the temperature of the saturator and cell. Data on water vapor pressure at various temperatures were taken from [25].

In the second case, the air consecutively flowed through the thermostated glass tube with drying agent (P_2O_5), saturator filled with a water – organic mixture and then through the sample cell. The temperature of the sample cell and saturator were 25°C. The water activity

(a_w) in vapor phase was adjusted by changing the water activity in the liquid water – organic mixture.

Water sorption by protein films was controlled in the region of OH stretching vibration band at 3450 cm^{-1}, which is the most intensive band of sorbed water.

Sorbed dioxane has some intensive vibrational bands. The bands at 2966 and 2862 cm^{-1} were assigned to the CH_2 asymmetric and symmetric stretching vibrations, respectively. The band at 1455 cm^{-1} was assigned to the CH_2 bending vibrations. Dioxane sorption by protein films was controlled at 1121 cm^{-1}, which is the most intensive band of sorbed dioxane. This band was assigned to the stretching vibrations of the C-O group. To test the reliability of the dioxane sorption measurements, the absorbance at 1121 cm^{-1} was correlated with the absorbance values at 2966, 2862 and 1455 cm^{-1}. Good linear correlations were observed between the relative absorbance values (D_x/D_o) at 2966, 2862 and 1455 cm^{-1} and the D_{1121}/D_o values for chymotrypsin [21].

Protein films were flushed by pure water vapour or water – organic vapour mixtures until no further changes were detected in the absorbance values at 3450 cm^{-1} (water sorption control) and 1121 cm^{-1} (dioxane sorption control).

Supposing that during sorption the protein film swells predominantly due to increasing thickness, the protein hydration and sorption of organic solvent vapour were calculated from the Eqs. 5 and 6:

$$h = \frac{2.303 \cdot S_{water}\, \varepsilon_P}{B_{water}\, D_P} \tag{5}$$

where h is the protein hydration, mol water mol^{-1} protein, S_{water} is the area of water absorbance band, cm^{-1}; ε_P is the protein molar extinction at the maximum of Amide I band, $Lmol^{-1}$ protein cm^{-1}; B_{water} is the water integral molar extinction coefficient, L mol^{-1} water cm^{-2}; D_P is the optical density at the maximum of amide I band. For pure water, it was taken that B_{water} = 96,000±1000 L mol^{-1} water cm^{-2} [20].

$$A = \frac{2.303 \cdot S_{org}\, \varepsilon_P}{B_{org}\, D_P} \tag{6}$$

where A is the sorption of organic solvent, mol organic solvent mol^{-1} protein, S_{org} is the area of organic solvent absorbance band, cm^{-1}; ε_P is the protein molar extinction at the maximum of Amide I band, L mol^{-1} enzyme cm^{-1}; D_P is the optical density at the maximum of amide I band.

B_{org} is the organic solvent integral molar extinction coefficient, L mol^{-1} cm^{-2}. The integral absorption extinction coefficient for dioxane (B_{org}) was calculated from the area of the absorption band of the pure substance in a cell with layer thickness of 10 μm: $B_{dioxane}$= 19,100±17 L mol^{-1} cm^{-2}.

The molar absorption extinction coefficient of HSA was determined measuring the amide I spectra of protein solutions in heavy water. The molar absorption extinction coefficient of human serum albumin (ε_P) is 237,600 ± 300 L mol^{-1} cm^{-1}. The molecular weight of human serum albumin was taken as 66,000 Da.

3. RESULTS AND DISCUSSION

3.1. Analysis of Organic Solvent Sorption

Figure 2 shows the dioxane sorption isotherm for human serum albumin at 25°C. The initial state of the protein film (zero hydration level) for studying the sorption of dioxane was obtained by drying in air at water activity less than 0.01. As shown in Figure 2, at low water activity, dioxane is virtually not adsorbed by the dried protein. At a_w~0.5, the dioxane adsorption increases sharply, attaining its maximum at a_w=0.7-0.8. At a_w>0.9, the adsorption decreases to nearly zero.

To generalise these findings, the dioxane sorption on human serum albumin was compared with the sorption of dioxane on another globular protein, bovine pancreatic α-chymotrypsin, in Figure 2. As concluded from Figure 2, a good agreement was observed between the dioxane sorption on HSA and α-chymotrypsin (CT). This agreement shows that the sorption of organic solvent does not depend on the function and secondary structure of a protein. The organic solvent sorption is proportional to the mass of protein sorbent and mainly determined by the protein primary structure. As shown in Table 1, the content of hydrophobic amino acid groups in the primary structure of HSA and α-chymotrypsin is close.

Table 1. The content of hydrophobic and polar amino acid groups in the primary structure of human serum albumin and bovine pancreatic α-chymotrypsin

Protein	Hydrophobic groups, %	Polar and ionizable groups, %
Bovine pancreatic α-chymotrypsin [26]	48.0	52.0
Human serum albumin [22]	41.5	58.5

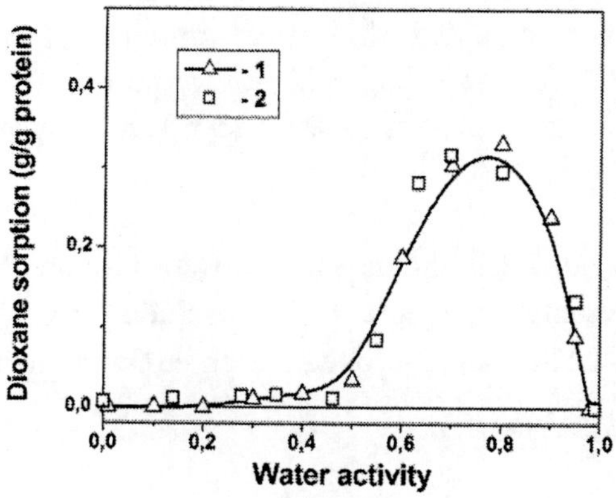

Figure 2. Sorption of dioxane by α-chymotrypsin and human serum albumin as a function of water activity. (1) Chymotrypsin. Modified data from [21]; (2) HSA. Modified data from [27].

3.2. Analysis of Water Sorption Isotherms

Figure 3 shows the water sorption isotherms for human serum albumin in the absence and presence of dioxane. The protein sample on which water was adsorbed was preliminary dried in air flow at a_w<0.01. As shown in Figure 3, the water sorption isotherms measured in the absence of organic solvent resemble typical sigmoid shaped curves. This fact is in accordance with the previous published results for serum albumin and other proteins [1, 2, 28].

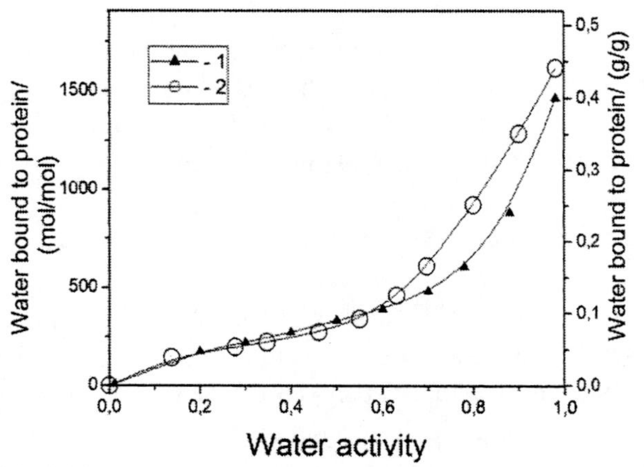

Figure 3. Water sorption isotherms for human serum albumin in the absence (1) and presence (2) of dioxane. Modified data from [27]. Solid lines were fitted by a set of polynomials.

The presence of organic solvent markedly affects the ability of the protein to bind water. The effect of dioxane molecules on the water sorption was characterised by the differences (Δh) in water uptake for the corresponding branches in the presence and absence of organic solvent vapor (Figure 4).

Two organic solvent effects on water sorption by human serum albumin were observed:

1. At low water activity (a_w<0.5), the water adsorption branch obtained in the presence of organic solvent is similar to that measured in the absence of dioxane (Figure 4).
2. At water activity more than 0.5, the water sorption isotherm in the presence of dioxane lies above the corresponding isotherm for pure water (Figure 4). This result corresponds to the organic solvent – assisted effect on water binding by human serum albumin.

To confirm the generality of these findings, the Δh values for another globular protein, α-chymotrypsin, were also presented in Figure 4. As concluded from Figure 4, the effect of dioxane on water binding by HSA is similar to that observed for α-chymotrypsin.

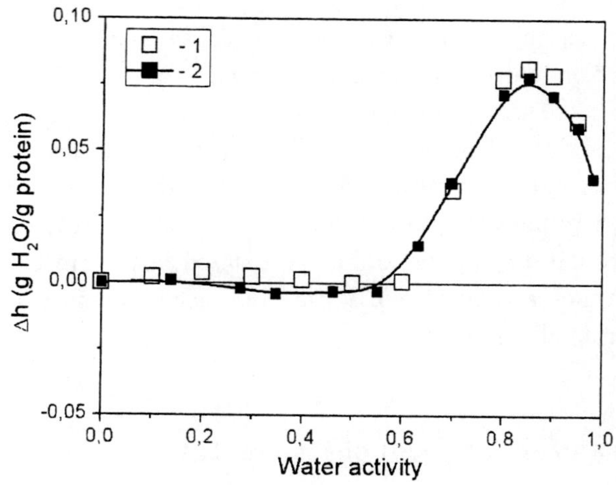

Figure 4. Difference (Δh) between the water sorption isotherms measured in presence and absence of dioxane: (1) Chymotrypsin; (2) HSA. Modified data from [21, 27].

The BET model [29] is widely used for describing the sorption ability of various solids, including proteins [1, 2, 28] (Eq. 7):

$$h = h_m \left[\frac{K a_w}{1 + K a_w} + \frac{a_w}{1 - a_w} \right] \qquad (7)$$

where h is the hydration of solid protein (mol water per mol protein). h_m is the number of water binding sites in the sorption monolayer (mol water per mol protein). This effective parameter charcterizes the number of sorption sites capable of firmly binding water molecules. K is the water sorption constant. This effective parameter characterizes the affinity of the protein for water.

It was found that the BET equation (Eq. 7) describes the water sorption isotherms displayed in Figure 3 up to a water activity no more than 0.6. This result is in close agreement with the data obtained in previous studies for serum albumin and other proteins [1, 2, 28].

Table 2. Parameters of water sorption by human serum albumin estimated by Eq. 7[A]

Type of sorption branch	h_m, mol water/mol protein	K	S_o[B]
	In the absence of dioxane		
	Human serum albumin		
Adsorption [28]	177.1 (1.4)	15.3 (1.3)	1.8
[27]	240.7 (3.4)	13.0 (0.9)	4.3
	In the presence of dioxane		
	Human serum albumin		
Adsorption [28]	168.9 (5.8)	13.3 (5.2)	14.6

The values of confidence interval of the parameters calculated by Eq. 9 are given in parentheses.
A) Applicable range (a_w): 0 - 0.6.
B) s_o is the residual standard deviation.
C) In [27], no adsorption-desorption cycle measurements were performed.

Water binding by serum albumin was determined in many studies. To confirm the reliability of our sorption data we also applied Eq. 7 to approximate the water vapor adsorption isotherm of horse serum albumin measured by Bull [28]. As concluded from Table 2, a good agreement is observed between the Bull's [28] and our data.

The water sorption parameters estimated from Eq. 7 are presented in Table 2. As seen from Table 2, the values of the water sorption constant, K, and the water sorption monolayer, h_m, for pure water sorption are similar to those measured in the presence of dioxane. This implies that, at low water activity, there is no significant organic solvent effect on the hydration of human serum albumin.

3.3. Analysis and Band Assignment of Protein Infrared Spectra

Infrared spectra of proteins contain some vibrational bands arising from the amide groups of protein backbone. Among them, the amide I band between 1700 and 1600 cm^{-1}, which is due primarily to the C=O stretch of the peptide linkages, is the most sensitive to changes in the secondary structure of proteins. The infrared spectra of human serum albumin in the amide I region are presented in Figures 5-8. Figures 5 and 6 show typical absorbance and second derivative spectra of human serum albumin in the absence of dioxane in the amide I region. The amide I contour consists of a number of bands at frequencies that are characteristic of specific types of secondary structure. It has been well documented that the frequency at maximum absorbance of the amide I band is determined by the predominant form of secondary structure in the protein [15-17]. The maximum absorbance of the amide I band of proteins with a high content of α-helix occurs between 1658 and 1654 cm^{-1} and with a high content of β-sheet structure the maximum is found between 1643 and 1632 cm^{-1}.

As shown in Figures 5 and 6, the most dominant band component of the HSA spectra is the band at 1654 cm^{-1}, which is usually attributed to α-helix structure. Upon hydration the infrared spectra of human serum albumin are grossly altered relative to that for the initial dried protein. There are increases in bandwidths and shifts in band positions and relative absorbances, which are indicative of protein conformational changes.

The hydration in the absence of dioxane markedly alters the secondary structure of human serum albumin as revealed by the changes in relative intensity at 1654 cm^{-1}. At low water activity ($h<0.1$ g/g protein), the hydration process induces pronounced structural rearrangements (Figure 9). The D_{1654}/D_0 curve reaches a plateau at water activity of 0.5 ($h>0.1$ g/g protein), indicating that the conformational changes are largely completed.

Figures 7 and 8 show the absorbance and second derivative spectra of HSA in the presence of dioxane in the amide I region. As seen from Figures 7 and 8, the spectra of HSA are very different from those in the absence of dioxane. Dioxane increases markedly the intensity of the 1654 cm^{-1} band.

The effect of dioxane on the HSA structure was characterized by changes in the relative intensity at 1654 cm^{-1}. The D_{1654}/D_o functions obtained in the presence of dioxane (Figure 9) have been compared with those determined for HSA in the absence of dioxane. This comparison can show any effects of the organic solvent molecules on the protein structure. The differences in the D_{1654}/D_o values obtained in the presence and absence of dioxane are presented in Figure 10. As concluded from Figure 10, the $\Delta D_{1654}/D_o$ function is consistent with the changes in the dioxane sorption by HSA (Figure 3). In the water activity range from 0 to 0.5, the $\Delta D_{1654}/D_o$ values are close to zero (Figure 10). Hence, it may be concluded, that the water-free dioxane has little effect on the structure of the initially dried protein. This fact agrees well with previously published results [30].

Above the threshold water activity of 0.5, the $\Delta D_{1654}/D_o$ values are sharply increased. The most significant effect of dioxane on the protein structure was observed at water activity of 0.7-0.8 (h>0.1 g/g protein) (Figure 10). It is well known that at a hydration level between 0 and 0.1 g/g, water is strongly bound to charged and polar protein groups [1, 2]. This implies that the primary hydration process (the hydration of polar and charged protein groups) is sufficient for the expression of the dioxane-induced structural rearrangements.

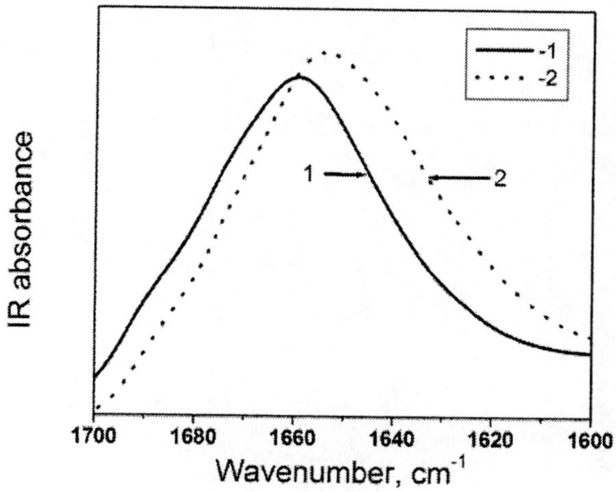

Figure 5. Typical absorbance spectra of human serum albumin in the amide I region in the absence of organic solvent. (1) a_w=0.01; (2) a_w=0.98.

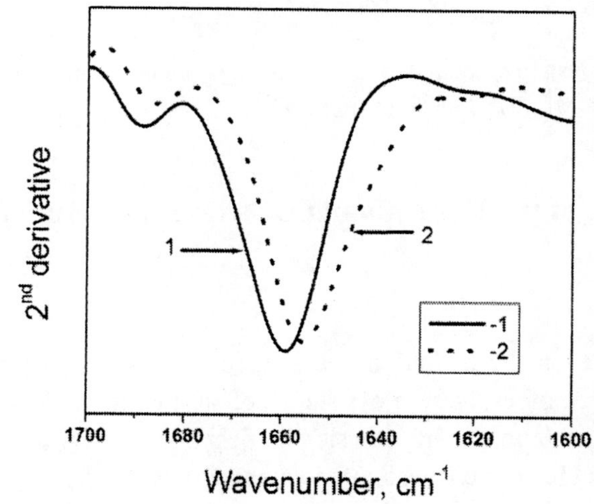

Figure 6. Typical second derivative spectra of human serum albumin in the amide I region in the absence of organic solvent. (1) a_w=0.01; (2) a_w=0.98.

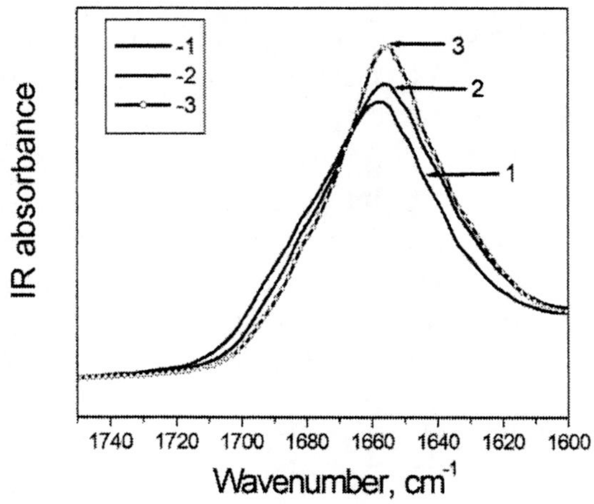

Figure 7. Typical absorbance spectra of human serum albumin in the amide I region in the presence of dioxane. (1) a_w=0.01; (2) a_w=0.27; (3) a_w=0.69.

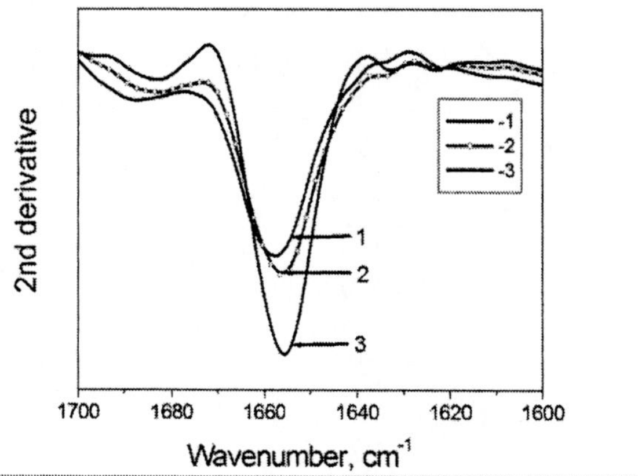

Figure 8. Typical second derivative spectra of human serum albumin in the amide I region in the presence of dioxane. (1) a_w=0.01; (2) a_w=0.27; (3) a_w=0.69.

3.4. Effect of Dioxane of the Hydration and Structure of Human Serum Albumin

Region 1 (a_w<0.5)

The effect of dioxane on the hydration and structure of human serum albumin can be interpreted based on the hypothesis of the exsistence of strong intermolecular contacts in the dried proteins [21]. According to this hypothesis, the dehydration of proteins is accompanied by the formation of intermolecular contacts, predominantly through hydrogen bindsand/or ion bridgesbetwen polar groups of the protein (carboxy, alcohol, amide, and amine groups). These processes result in the formation of a glassy-like structure with an enchanced rigity that is substantially distorted compared to the native structure. In this case, being involved in the formation of interprotein contacts, polar/charged groups in a dry protein are incapable of functioning as sorption sites.

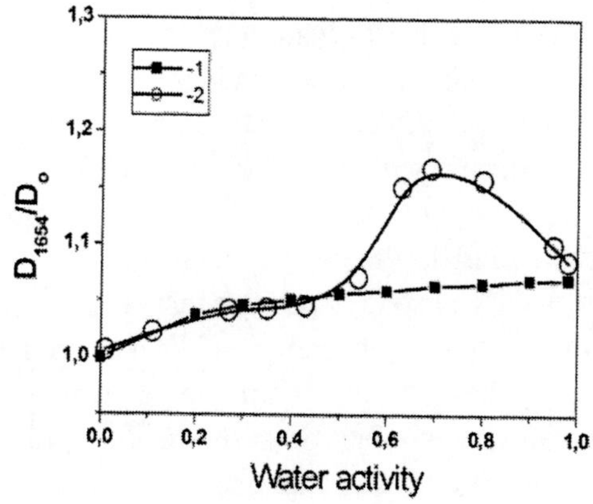

Figure 9. Human serum albumin. Absorbance at 1654 cm^{-1} as a function of water activity: (1) In the absence of dioxane; (2) In the presence of dioxane. D_o is the absorbance value at 1654 cm^{-1} for human serum albumin in the absence of dioxane at a_w=0.01.

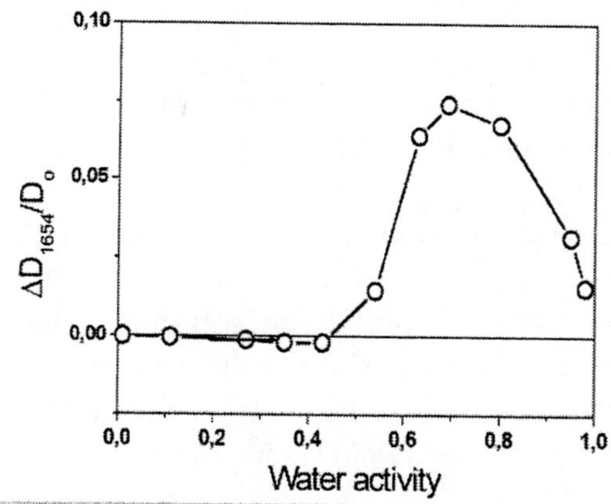

Figure 10. Difference in the absorbance at 1654 cm^{-1} as a function of water activity. D_o is the absorbance value at 1654 cm^{-1} for human serum albumin in the absence of dioxane at a_w=0.01.

On the other hand, it was demonstrated [30] that the stability and structure of dry proteins are crucially dependent on the ability of the organic solvent to form hydrogen bonds. Only solvents capable of forming strong hydrogen bonds (for example, DMSO) produce appreciable structural rearrangements and heat effects. Thus, when forecasting the the effect of an organic solvent on the sorption and structural properties of a protein, it is necessary to evaluate the tradeoff between the proton-donor and proton-acceptor properties of the solvent. It is an important aspect, because when an H-bonded contact is severed, sorbate molecules (those of water or dioxane) have to choose to join either a proton-donor or proton-acceptor fragment of the broken contact.

Water molecules will solvate both the proton-donor and proton-acceptor groups of human serum albumin, while dioxane molecules will solvate only the proton-donor group of the broken contact. The remaining proton-acceptor group will be solvated by water molecules. Thus, dioxane molecules alone, in the absence of water molecules, cannot make interprotein contacts disengage. In addition, dioxane molecules cannot be sorbed jointly with water molecules, probably due to steric and diffusional limitations. The molar volume of dioxane

(85.2 cm^3/mol) is more than four times that of water (18 cm^3/mol). Therefore, dioxane is not appreciably adsorbed at rather low water activities. When a dry protein was introduced into water-free dioxane, no significant heat effects and structural rearrangements were observed [30]. Thus, the main process that occurs within a_w = 0-0.5 is the interaction of water with the protein.

According to the proposed model, this stage of sorption is the insertion of water molecules into the structure of the dry protein, a process that is accompanied by the rupture of interprotein contacts, hydration of the polar groups of broken contacts, and formation of sorption sites. An alternative mechanism, the interaction between sorbate molecules and sorption sites without change in the number of sorption sites, also cannot be excluded, although, as an auxiliary one, since it does not explain why new dioxane sorption sites begin to intensively formed at a_w = 0.5. All these findings, taken into together, make a compelling case that the stage of rupture of interprotein contacts strongly affects the properties of dry proteins.

Region 2 (a_w > 0.5)

The sorption of dioxane occurs intensively at a_w >0.5, after the rupture of most the interprotein contacts. Water molecules attach to both proton-donor and proton-acceptor fragments broken contacts, while dioxane molecules compete with water molecules onle for proton-donor fragments.

On the other hand, each dioxane molecule contains four hydrophobic methylene groups. Therefore, it is reasonable to assume that, when inserted into the structure of a hydrated protein sorbent, dioxane molecules are capable of solvating polar groups not only in hydrophilic but also in hydrophobic regions of the protein macromolecule, thereby creating new sorption sites for water molecules. Dioxane produces an activating effect on the protein, making it adsorb an additional amount of water (Figure 4). The maximum at water activities reflects the competition for additional sorption sites on the protein between water and dioxane.

The propose model suggests that, at high water activity, the behaviour of the system is determined by the interplay of two factors:

i. The competition between water and dioxane for new sorption sites formed as a result of the rupture of interportein contacts and
ii. The activating effect of dioxane on water sorption. The maxima in the water activity dependencies of the dioxane sorption and excess amount of water sorbed (compared to the sorption of pure water) reflect the competition between these factors.

CONCLUSION

1. FTIR spectroscopy provides an effective and selective tool for studying the hydration of proteins in the absence and presence of organic solvents and the corresponding conformational rearrangements in the whole range of water activity.

2. The results obtained demonstrate that the hydration and structure of human serum albumin depend markedly on how the protein has been hydrated – whether in the presence or in the absence of organic solvent.
3. The H-donating and H-accepting properties of organic solvents are important for estimating the possible effect of organic molecules on the structure and hydration of proteins.

REFERENCES

[1] Gregory, R.B. (1995) Protein hydration and glass transition behavior. In: *Protein-Solvent Interactions (*Gregory, R.B., ed.*)*, Marcel Dekker, New York, 191-264.

[2] Rupley, J.A., Careri, G. (1991) Protein hydration and function. *Adv. Protein Chem,* 41, 37-172.

[3] Klibanov, A.M. (2001) Improving enzymes by using them in organic solvents, *Nature,* 409, 241-246.

[4] Carrea, G., Riva, S. (2000) Properties and synthetic applications of enzymes in organic solvents, *Angew. Chem. Int. Ed.,* 39, 2226-2254.

[5] Gupta, M.N. (2000) *Methods in Non-Aqueous Enzymology*, (M.N. Gupta, ed.), Birhäuser Verlag, Basel-Boston-Berlin.

[6] Ke, T., Klibanov, A.M. (1998) On enzymatic activity in organic solvents as a function of enzyme history, *Biotech. Bioeng. 57, 746-750.*

[7] Halling, P.J. (2004) What can we learn by studying enzymes in nonaqueous media? *Philos. Trans. R. Soc. Lond. B Biol. Sci.,* 359, 1287– 1297.

[8] Partridge, J., Hutcheon, G.A., Moore, B.D., Halling, P.J. (1996) Exploiting hydration hysteresis for high activity of cross-linked subtilisin crystals in acetonitrile, *J. Amer. Chem. Soc.* 118, 12873-12877.

[9] Zaks, A., Klibanov, A.M. (1988) The effect of water on enzyme action in organic media, *J. Biol. Chem.* 263, 8017-8021.

[10] Yamane, T., Kojima, Y., Ichiryu, T., Shimizu, S. (1988) Biocatalysis in a microaqueous organic solvent, *Ann. NY Acad. Sci.* 542, 282-293.

[11] Lee, C.S., Ru, M.T., Haake, M., Dordick, J.S., Reimer, J.A., Clark, D.S. (1998) Multinuclear NMR study of enzyme hydration in an organic solvent, *Biotechnol. Bioeng.* 57, 686-693.

[12] McMinn, J.H., Sowa, M.J., Charnick, S.B., Paulaitis, M.E. (1993) The hydration of proteins in nearly anhydrous organic solvent suspensions, *Biopolymers* 33, 1213-1224.

[13] Parker, M.C., Moore, B.D., Blacker, A.J. (1995) Measuring enzyme hydration in nonpolar organic solvents using NMR, *Biotechnol. Bioeng*. 46, 452-458.

[14] Halling, P.J. (1990) High-affinity binding of water by proteins is similar in air and in organic solvents, *Biochim. Biophys. Acta* 1040, 225-228.

[15] Prestrelsky S.J., Tedeschi N., Arakawa T., Carpenter J.F. (1993) Dehydration-induced conformational transitions in proteins and their inhibition by stabilizers. *Biophys. J,* 65, 661-671.

[16] Van de Weert M., Harvez P.I., Hennink W.E., Crommelin D.J.A. (2001) Fourier transform infrared spectrometric analysis of protein conformation: effect of sampling and stress factors. *Anal. Biochem,* 297, 160-169.

[17] Griebenow K., Klibanov A.M. (1995) Lyophilization-induced reversible changes in the secondary structure of proteins. *Proc. Natl. Acad. Sci. USA,* 92, 10969-10976.

[18] Sirotkin, V.A., Zinatullin, A.N. Solomonov,B.N., Faizullin, D.A., Fedotov, V.D. (2002) Interaction enthalpies of solid bovine pancreatic chymotrypsin with organic solvents: comparison with FTIR-spectroscopic data. *Thermochim. Acta*, 382, 151-160.

[19] Sirotkin V.A., Korolev D.V. (2005) Effect of acetonitrile on the hydration of human serum albumin films: a calorimetric and spectroscopic study. *Thermochim. Acta,* 432, 246-253.

[20] Sirotkin V.A., Solomonov B.N., Faizullin D.A., Fedotov V.D. (2002) Sorption of water and acetonitrile by human serum albumin. *Russ. J. Phys. Chem. A,* 76, 2051-2057.

[21] Sirotkin, V.A. (2005) Effect of dioxane on the structure and hydration-dehydration of α-chymotrypsin as measured by FTIR spectroscopy. *Biochim. Biophys. Acta.*, 1750, 17-29.

[22] Carter, D.C.; Ho, J.X. (1994) Structure of serum albumin, *Adv. Prot. Chem. 45*, 153-203.

[23] Perrin, D.D., Armarego, W.L.F., Perrin, D.R. (1980) *Purification of Laboratory Chemicals*, Oxford: Pergamon Press.

[24] Kogan, V.B., Fridman, B.N., Kafarov, V.V. (1966) *Equilibrium between Liquid and Vapor. Handbook*, Nauka, Moscow.

[25] Nikol'skii, B. P. (1963). *Chemist's Handbook,* Goskhimizdat: Leningrad, Russia.

[26] Meloun, B., Kluh, I., Kostka, V., Moravek, L., Prusik, Z., Vanasek, J., Keil, B., Sorm, F. Covalent structure of bovine chymotrypsinogen A, *Biochim. Biophys. Acta.* 130 (1966) 543-546.

[27] Sirotkin, V.A., Faizullin, D.A. (2004) Interaction enthalpies of solid human serum albumin with water-dioxane mixtures: comparison with water and organic solvent vapor sorption, *Thermochim. Acta*, 415, 127-133.

[28] Bull, H.B. (1944) Adsorption of water vapor by proteins. *J. Amer. Chem. Soc.* 66, 1499-1507.

[29] Brunauer, S., Emmet, P.H., Teller, E. (1938) Adsorption of gases in multimolecular layers, *J. Amer. Chem. Soc.* 60, 309-319.

[30] Sirotkin, V.A., Zinatullin, A.N. Solomonov,B.N., Faizullin, D.A., Fedotov, V.D. (2001) Calorimetric and Fourier transform infrared spectroscopic study of solid proteins immersed in low water organic solvents. *Biochim. Biophys. Acta.* 1547, 359-369.

In: Human Serum Albumin (HSA)
Editor: Travis Stokes
ISBN: 978-1-63482-963-2

Chapter 7

CAPIDAN, A FLUORESCENT REPORTER FOR DETECTION OF ALBUMIN DRUG-BINDING SITE CHANGES

***G. E. Dobretsov*[1]*, T. I. Syrejschikova*[2]*, N. V. Smolina*[1] *and M. G. Uzbekov*[3]**

[1]Research Institute for Physical Chemical Medicine, Moscow, Russia
[2]Lebedev Physical Institute, Moscow, Russia
[3]Moscow Research Institute of Psychiatry, Russia

ABSTRACT

CAPIDAN, a special fluorescent dye, binds to drug binding sites of human serum albumin. About 90-95 per cent of its fluorescence in serum or plasma originates from albumin-bound dye. This specificity opens a new way to study albumin in unfractionated serum saving intact albumin structure. Due to these properties, CAPIDAN has shown that a variety of diseases is accompanied by some changes of the drug-binding sites. This is a new source of information useful for clinics. That sensitivity to site abnormalities is raised by using time-resolved measurements of fluorescence decay and three new methods of the data analysis: (1) estimation of absolute values of concentration of fluorescent species in heterogeneous biological objects; (2) measurement of dye rotational mobility in heterogeneous albumin binding sites; and (3) method of chloride/nitrate substitution to detect interaction of a drug negative charge with arginine positive charges in the albumin binding sites.

1. INTRODUCTION

Measurement of albumin mass concentration in serum is obligatory assay in clinical practice. It is commonly clear that low level of albumin is a negative factor in disease prognosis and is associated with low lifespan. On the other hand, it has been shown that, besides quantity, albumin "quality", its capacity for binding and transferring a lot of molecules (different metabolites, fatty acids, hormones, xenobiotics including drugs etc), its ability to clean the body changes due to state of body, and these changes could be very important.

In 1925, it has been found out that a dye, phenol red, binds to albumin much better than to other plasma proteins [1]. Further studies have shown that albumin has much more ability to bind a lot of dyes as well as drugs than other plasma proteins. As a result, human serum albumin (HSA) plays an important role in pharmacokinetics [2-4].

Interest to molecular mechanisms of drug-albumin interaction has raised after the Second World War while it has been found out that penicillin is transported in blood by HSA [3, 5]. In these studies it was shown that the interaction is not covalent, it is reversible, and that drugs and dyes can occupy the same HSA sites competitively. However dyes are quite convenient for that research because their absorption spectra change significantly at binding with HSA [6-8]. Moreover, some dyes including a bromocresol group [9] prefer to bind with HSA in whole plasma (or serum) in spite of the presence of many other proteins. This observation allowed proposing a clinical method of HSA assay which is used today worldwide [10].

However as early as in the 1950-1960 some abnormalities of dye binding have been observed in various diseases. A few researchers started investigation of these deviations. One of the dyes used on the early steps of the research was congo red. Its ability to bind to plasma proteins has been known for a long time (Sorensen 1909: citation from [11]). Csögör has proposed a technique to measure HSA-bound congo red and has shown that its quantity is disease-sensitive [11]. Some other researchers used this technique and confirmed sensitivity of congo red test [12] as well as several other dyes (e.g., [13]). Unfortunately, these methods were rather laborious, therefore systematic clinical studies were not performed.

In 1952, Laurence and Weber have studied interaction of various fluorescent dyes with proteins. An aniline dye, N-phenyl-1-amino-8-sulphonaphthalene (ANS), had very low fluorescence in water, but its fluorescence intensity raised many times in protein, especially albumin solutions [14, 15]. In clinical samples of sera, ANS fluorescence intensity was roughly proportional to HSA concentration but in some samples, especially of jaundice patients, the intensity was decreased [16]. These results had no continuation. Ten years later ANS has been used as a fluorescent probe for studying bovine serum albumin [17-20] and some other proteins [21]. It has been shown that bovine serum albumin has 15 or more binding sites for ANS, but only 5 of them have rather high binding constant of the order of $1 \cdot 10^6$ M^{-1} [22].

Following ANS, other fluorescent dyes have been tested to detect HSA changes at pathological processes (e.g., see review [23]). These HSA abnormalities are of significant interest. Therefore a new fluorescent probe named CAPIDAN (or K-35) has been synthesized by B.M.Krasovitsky and colleagues especially for study of HSA pathological changes.

In this chapter, we are focused mainly on physical aspects of CAPIDAN-HSA interaction; a brief review of its clinical testing is proposed also.

2. CAPIDAN: Clinical Data

Human serum albumin represents about 60% of mass of human serum proteins. Its half-life is approximately 20 days [24]. HSA is synthesized in liver [25]. Albumin migrates from plasma to tissues and back, its plasma pool is replaced by the tissue pool during 24 hours [26]. Diseases can upset circulation/distribution of HSA in the body (hepatitis, heart failure, nephrosis) [27, 28].

HSA has a lot of important functions: maintenance of the osmotic pressure [24] and acid/ base balance [24; 29], regulation of vascular permeability and hemostasis [30, 31]. HSA can regulate apoptosis [33, 34] and this process depends on the state of HSA thiol groups and presence of some ligands including fatty acids [35]. Albumin is an effective free radical scavenger (both oxygen and nitrogen species) [36].

Another important function of HSA is binding and transport of low-weight molecules (naturally occurring substances: fatty acids [37], bile pigments [38], cholesterol [39], nitrogen oxide [40], cations Zn^{2+}, Ca^{2+}, Ni^{2+}, Cu^{2+}, Fe^{2+} [41, 42] and xenobiotics, particularly drugs. Interaction of HSA with free radicals can lead to alteration of albumin binding capacity towards other ligands, including drug substances [43, 44].

Due to binding to HSA, chemical and therefore biological activity of ligand changes. For example, binding of transition valency metals to HSA causes some hundredfold decrease of their activity and capacity to generate radical. Many substances are toxic in free state but after binding to albumin their toxicity significantly lowers [45]. HSA transports these toxicants mainly to the liver where they degrade. Consequently HSA plays quite important role in detoxification system of the body.

Binding of a ligand to albumin molecule occurs in special binding sites. Most endogenous substances have their own special sites (bile and fatty acids, bilirubin, divalent ions). Majority of drugs bind to drug sites I and II. Often competition of different ligands for the same binding place occurs, therefore the actual toxicity of the displaced ligand can increase. Changes of HSA binding capacity in different diseases can enhance toxic effects of drugs and metabolites. Abnormality of albumin distribution between blood stream and other tissues has more pronounced effect on drug pharmacokinetics then simple reduction of albumin concentration [46]. Therefore study of HSA properties particularly of its binding capacity is essential in developing of new drugs.

Albumin Fluorescence Assay and Its Clinical Application

HSA binds different dyes including fluorescent probes.

The fluorescent probe CAPIDAN (N-carboxyphenylimide of dimethylaminonaphthalic acid) was synthesized and kindly gifted by professor B.M.Krasovitsky and colleagues (Institute of Monocrystals, Kharkow, Ukraine). The probe binds to albumin molecule and selectively fluoresces from albumin fraction in the whole serum or plasma [47]. The approximative localization of CAPIDAN in albumin globule is near the Sudlow's drug-binding site I (see section 5).

A kit to study HSA in its native state (avoiding the fractionating or any treatment of plasma or serum) was developed due to CAPIDAN selective fluorescence. The kit measures two values:

1) effective albumin concentration (EAC) characterizing the quantity as well as the state of binding sites,
2) total albumin concentration (TAC) equal to albumin concentration assayed in clinical laboratories with bromcresol green.

To measure EAC, it is necessary to add the abundance of CAPIDAN to the diluted serum or plasma at pH 7.4. Fluorescence intensity of CAPIDAN under these conditions depends on the state of binding sites which is close to their state in the native plasma without any treatment, on presence of drugs in the "drug" sites or bilirubin or fatty acids etc. In order to determine TAC, we adjusted pH to 4.2 in the same sample and added detergent Brij-35. This procedure provokes "N-F"-transition in HSA globules [48]. Then CAPIDAN fluorescence intensity is approximately proportional to the albumin content and can be converted into the albumin concentration units (g/l). It is considered that in norm EAC comes close to TAC, their ratio EAC/TAC = 1 ± 0.05. EAC/TAC doesn't depend on concentration and reflects only changes of binding sites and the molecule conformation. Based on these parameters the specific coefficient of HSA binding capacity, "reserve of albumin binding capacity" (RAB), can be calculated according to the following formula: RAB = EAC/TAC x 100%. RAB can be assumed as a measure for a fraction of functionally active HSA, in per cent. However RAB reflects both the binding ability and structural changes of HSA sites.

Numerous studies using this HSA fluorescence assay have shown that decrease of EAC and EAC/TAC accompanies different diseases. The fact has led to the hypothesis that changes of HSA conformation are a special response of the body to pathological processes [49].

Often changes of HSA correlate with the disease form or severity. So the estimation florescent HSA factors can be useful for widening of diagnostics informativity and for the control of treatment efficiency.

Often changes of HSA correlate with the disease form or severity. So, the fluorescent detection of these HSA changes can be useful for widening of diagnostics informativity and for the control of treatment efficiency.

The List of Diseases and Pathological States Which Shown CAPIDAN to Be Quite Informative Is Rather Long

- surgical diseases and critical conditions (peritonitis [50], appendicitis [51], acute pancreatitis, pancreatonecrosis [52-54], burn disease [55], chemical poisoning [56], abscess [57], hernias [58], multitrauma, nurotrauma and critical conditions [59-63]);
- internal diseases (miocardial infarction [64, 65], cardiac failure [66], cerebral ischemia [67], ischemic stroke [68], pneumonia and chronic obstruction lung diseases, hepatitis and hepatic cirrhosis [69], renal failure [70], haemodyalysis, nephropathy, urinemia [71];
- ophtalmology (glaucoma [72], senele cataract [73] pseudoexfoliation syndrome);
- children diseases and neonatology (multiorgan failure in newborns, kernicterus [74];

- mental disorders (schizophrenia [75-77], depressive syndrome [78-80], and acute poisoning with psychotic drugs [81-83].

Some results on the usage of the assay or its modifications are described below in more details.

Myocardial Infraction

In the begging of the nineties it was shown that immediately after the onset of noncomplicated transmural infraction EAC was normal and during two days decreased for 30%. TAC remained unchanged for two days and lowered for 20% during the next days. In lethal form both factors decreased significantly. In jugulated angor pectoris only EAC lowered but returned to normal values in two days [64].

In [65] the dynamics of EAC and TAC in acute infarction was observed for three weeks. 30% reduction of EAC was found after the onset of the disease and stayed lowered till the end of the monitoring. TAC didn't vary significantly during the period. Therefore there were little alterations of HSA concentration in acute cardiovascular disorders but CAPIDAN showed conformational abnormalities that occurred during at least three weeks. EAC and EAC/TAC correlated with the severity and form of the disease. As it is shown below (section 10), CAPIDAN fluorescence falls at decrease of HSA-bound fatty acids. Probably these infarction-accompanied changes of HSA properties relate to the abnormalities or failure of fatty acid transportation by HSA.

Acute Pancreatitis

Measuring of EAC and TAC can be helpful both to differentiate edematous and necrotising pancreatitis [52] and predict the progression of the disease [54].

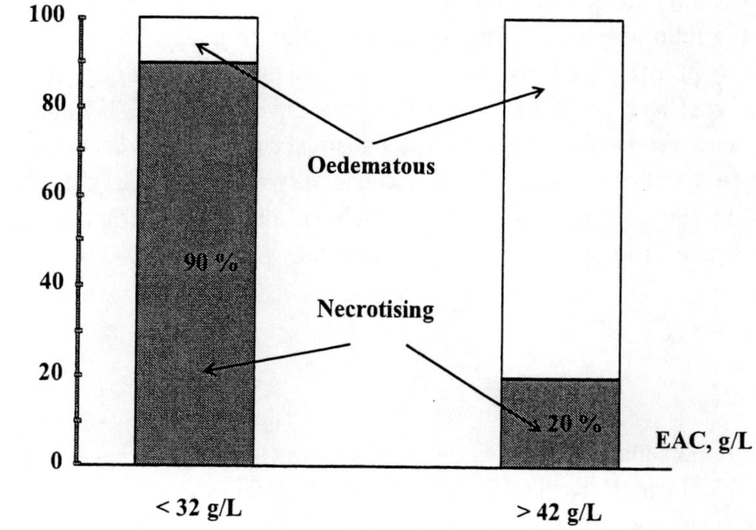

Figure 1. CAPIDAN fluorescence from HSA (effective albumin concentration, EAC) differentiates oedematous and necrotising pancreatitis on the first day of admission.

EAC more than 42 g/l at the first day of the admission gave the probability of necrotic form 20%. If EAC was less than 32 g/l the probability of necrotizing pancreatitis grews towards 90% (Figure 1). And even a singular decrease of EAC (despite subsequent restoration of the parameter in some cases) during the period of the disease to 14 g/l means the significant raise of the death probability [54].

Diffuse Peritonitis

The efficiency of clinical scores for outcome of diffuse peritonitis is not sufficient and the mortality rate in peritonitis remains high. More than 200 patients were examined in a multicentre study [50, 60] and EAC and TAC were compared with results of routine analysis and SAPS and MPI scales. The combination of two parameters – EAC and urea concentration – at the first day of admission was found to be the most useful for outcome prognosis, the efficiency of this combination was very similar to the SAPS data. 95% of benign outcome of diffuse peritonitis followed blood urea level less than 8 mM and EAC more than 14 g/l. High urea (> 8 mM) in combination with low EAC (< 14 g/l) led to benign outcome only in 10% of cases. However the probability of benign outcome was 50% if EAC was between 14 and 22 g/l despite high urea concentration. TAC was less useful for prognosis of diffuse peritonitis. More pronounced alteration of EAC is due to both albumin migration to exudate and tissues and changes of binding sites properties. This can help in an early individual prognosis of diffuse peritonitis.

Peritoneal Exudation at Peritonitis and Acute Pancreatitis

In these cases an inflammation in abdominal cavity leads to penetration of plasma proteins including serum albumin from blood into the cavity. The fluorescent method of the total albumin concentration (TAC) assay was used to estimate relationship between albumin concentration in plasma (C_P) and in peritoneal exudate (C_E) [84]. These concentrations have been measured in dynamics of the inflammatory process during several days after admission. In this study it was often detected a significant negative correlation of C_P and the ratio C_E / C_P ; i.e., the more the ratio, the lower albumin concentration in plasma. Some examples of this correlation are shown in Figure 2. About half of all patients have this negative correlation at rather high linear coefficient $|r| > 0.7$; 14 % of patients have r between –0.7 and –0.3.

This strong tendency needs to have a clear statistical support. However it is difficult to do it simply, i.e., by a direct combining all data because of wide variation of personal data. In the patient group of the high negative correlation (51 % of all patients) a linear equation can be approximately used: for i-th patient in j-th measurement

$$(C_P)_{ij} \approx a_i + b_i (C_E / C_P)_{ij} \tag{1}$$

where a_i and b_i are personal coefficients. a_i vary between 16 and 103, b_i vary between –3 and –65. These individual variations significantly lower the total correlation coefficient. Therefore a procedure has been proposed to unify these data. Step 1: to calculate mean a_i and mean b_i for m patients:

$$a = \sum_{i=1}^{i=m} a_i = 47; \qquad b = \sum_{i=1}^{i=m} b_i = -23 \tag{2}$$

Step 2: each individual a_i is normalized to the mean a , i.e., shifts of intercepts along Y axis are averaged; each individual b_i is normalized to the mean slope of straight lines (see Figure 2), i.e., to b . A normalized function $f(C_P)_{ij}$ is introduced instead of $(C_P)_{ij}$:

$$f(C_P)_{ij} = a_i - (b / b_i)a_i + (b / b_i)(C_P)_{ij} \tag{3}$$

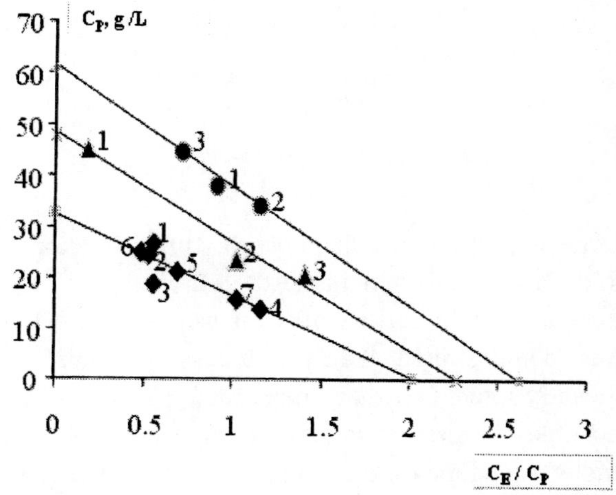

Figure 2. Three examples of patients with high correlation of plasma HSA concentration C_P and the ratio Exudate/Plasma, C_E /C_P. Dynamics day-by-day is shown by numbers near experimental points.

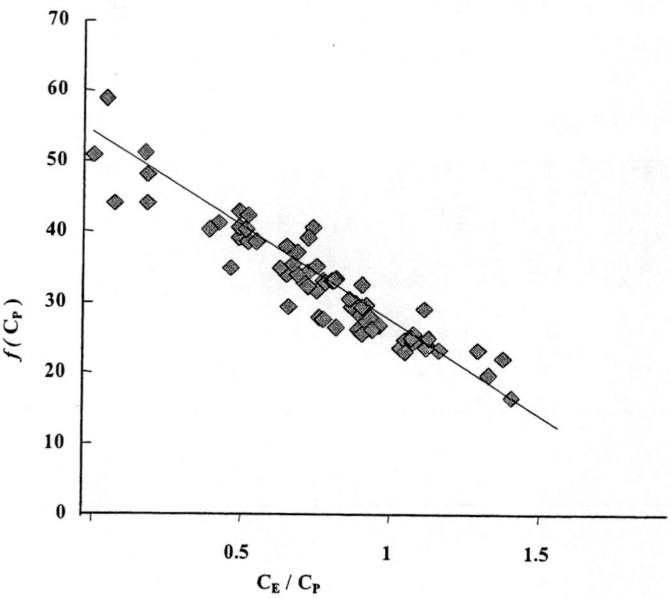

Figure 3. Linear correlation of the function $f(C_P)_{ij}$ and the ratio C_E/C_P in group of patients with personal r_i between –0.7 and –1. The total r = –0.94 ±0.05 at $p< 0.01$.

Step 3: a linear correlation coefficient between $f(C_P)_{ij}$ and the ratio C_E/C_P is calculated. This transformation does not influence personal coefficients r_i , but allows to do the total tendency more clear. Indeed, each patient in this group has the personal r_i between –0.7 and –1, however the total r after that transformation for all this group is very high: it is equal to r = –0.94 ±0.05 at p< 0.01. So, the tendency of high linear relationship between the ratio (albumin in exudate)/(albumin in plasma) and albumin in plasma is supported by the high correlation coefficient [84]. These relationship can be described well (Figure 3) by the equation:

$$f(C_P)_{ij} \approx a + b(C_E / C_P)_{ij} \quad (4)$$

Psychiatric Disorders

Shizophrenia

There is an urgent need in tests for diagnostics, differentiation (distinguishing) and prognosis of psychiatric disorders and to estimate the efficiency of psychopharmacotherapy or other therapeutic interventions. In early works on using the CAPIDAN as a probe for clinical purposes it was demonstrated that properties of albumin globule (EAC and EAC/TAC) were significantly altered in these disorders [85, 86]. But the most interesting results were obtained in a study on first-episode drug-naive schizophrenic (FES) patients due to absence of the previous psychopharmacotherapy and thus keeping intactness of HSA binding sites due to loading with antipsychotics and to integrity of methabolism which hadn't undergone pathological changes yet [87]. Steady-state fluorescence analysis didn't reveal any changes of HSA site properties [88]. However some modifications of the test medium and analysis of CAPIDAN fluorescence decay showed the significant difference between state of

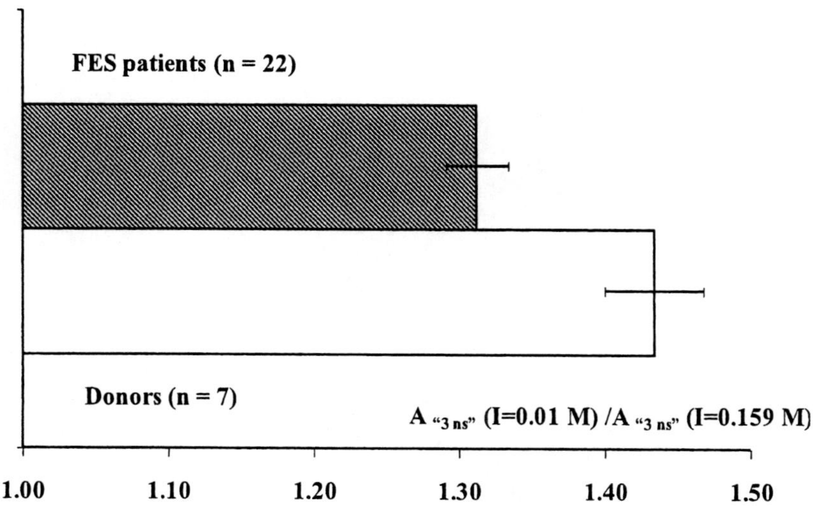

Figure 4. Effect of change of medium ionic strength on amplitude of the component with decay time 3 ns in donors and first-episode drug-naive schizophrenic patients, p < 0.001. The data are given as M ± m. Horizontal axis – ratio of the amplitude at low and physiological ionic strength.

HSA binding sites of the FES patients (ICD-10: paranoid (F20.0) and undifferentiated (F20.3) schizophrenia) and healthy people. CAPIDAN fluorescence decay is complex and has three components with decay time 9 ns, 3 ns and 1 ns (in detail see below section 5). The change of ionic strength of the test medium had different effect on the probe fluorescence decay component with decay time 3 ns in donors or schizophrenics, $p < 0.001$ [89, 90] (Figure 4).

Anxious Depression

As it was shown earlier for schizophrenia (see the previous section), CAPIDAN demonstrated difference between HSA properties of healthy individuals and long-term depression patients [79, 91]. As there is no techniques for early evaluation of the efficiency of therapy of depression and individualization of psychopharmacological treatment can be long enough, HSA fluorescent test has been used for the purpose [80]. 10 depressive patients (F32.1 (depressive episode as an independent disease) according to ICD-10) were under observation for tree – four weeks of psychopharmacological treatment. There were no any therapeutic intervention before the begging of the study. CAPIDAN fluorescence decay analysis (in details see below in section 5) revealed that mean fluorescence intensity of the probe in patient HSA before the beginning of the treatment is lower than in healthy donors, $p < 0.01$. CAPIDAN fluorescence decay is complex and has three components with decay time 9 ns, 3 ns and 1 ns (see below, section 5). Figure 5 shows amplitudes of the 9 ns decay component of CAPIDAN in patients' HSA before and within three weeks of psychopharmacotherapy. Almost all patients showed an increase of amplitude "9 ns" after the treatment ($p<0.01$ according to Wilcoxon's criterion for coupled variants). Mean raise of amplitude "9 ns" after the treatment was 15%. So HSA fluorescent data is rather sensitive to pathological process in anxious depression and can reflect HSA changes due to psychopharmacological treatment.

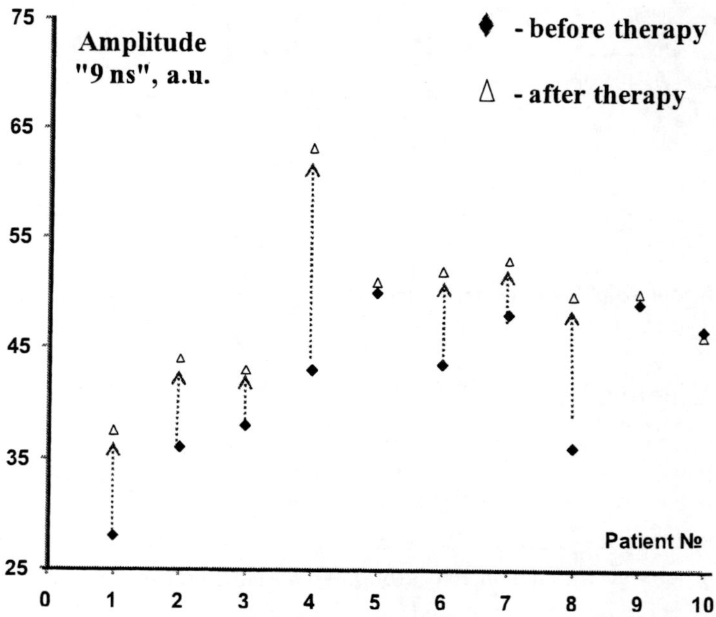

Figure 5. Change of amplitude "9 ns" of CAPIDAN fluorescence decay in patients with anxious depression after three weeks therapy. Amplitude is expressed in arbitrary units.

Acute Poisoning with Psychotropics and Its Complications

HSA fluorescence test (a steady-state version, see below, section 4) was compared to usual laboratory tests for endotoxemia (a hematology and biochemistry panel, middle molecules) in patients (n=513) with acute overdose with psychotropics [81]. On the first day of admission HSA fluorescence test was the best to predict lethal cases in the patients with medium and severe poisoning. Pneumonias often complicates the disease and results in 60% mortality. Clinical evidence of the pneumonia is opaquing, particularly at the beginning of the disease, that complicates pneumonia prognosis. CAPIDAN fluorescence intensity in HSA of the patients with following pneumonia is often lower than in patients without pneumonia. This difference helps to select a group of subjects just having or at risk of following [82].

Thus HSA fluorescence test with CAPIDAN and its modifications can be of prognostic and diagnostic value in many pathologies.

3. CAPIDAN MOLECULE

CAPIDAN molecule is shown in Figure 6. It contains a carboxylic group which losses H+ in water solutions; its pK_a is near 4.2, therefore CAPIDAN molecule is negatively charged at pH 7. As it is known, negatively charged dyes bind to HSA better than uncharged and positively charged ones [7, 8]. This CAPIDAN carboxyl should be localized on the boundary "binding site/water phase" due to hydration the negative charge by water.

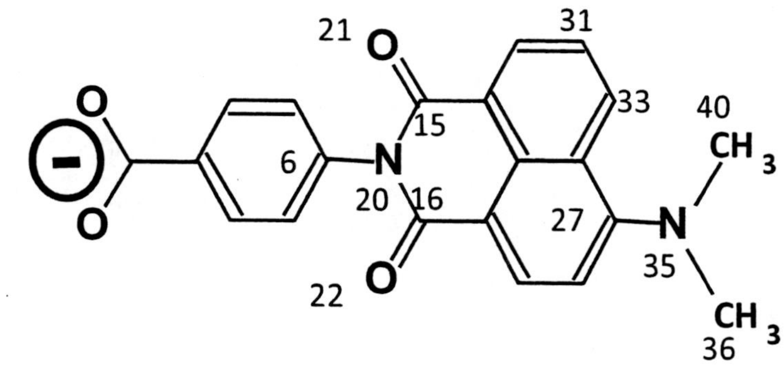

Figure 6. CAPIDAN molecule. Ionic form in water solutions at pH 7. Numeration of atoms is shown.

3.1. Quantum Chemical Data

Bond lengths and atom charges have been calculated recently in solvents of different polarity [92]. As can be seen, CAPIDAN molecule is not nonpolar and hydrophobic. Intermolecular charge shift (Table 1) rises creates dipole moments. Two important moments are originated from charge transfer in two groups O=C–N; therefore one can expect that this fragment of the molecule has to be solvated by polar environment. The charge shift along the long axis of the molecule is responsible for a general dipole moment of about 10 D. This moment is very important for absorption and fluorescent spectra shifts in polar media as well as in HSA (details see below). These three moments are shown in Figure 7.

All these charge shifts raise in polar environment (Table 1).

Table 1. CAPIDAN atom charges as natural bond orbital (NBO) in solvents of different polarity: computer simulation. ε is the dielectric constant. For details see [92]

Solvent	Heptane	Acetone	Water
Solvent polarity ε	1.9	22	80
Atom	Charge		
C_6	0.21	0.20	0.20
N_{20}	0.08	0.14	0.15
C_{15}	0.41	0.49	0.52
O_{21}	–0.70	–0.73	–0.74
C_{16}	0.63	0.71	0.74
O_{22}	–0.70	–0.74	–0.74
C_{27}	–0.59	–0.56	–0.56
N_{35}	–0.15	–0.17	–0.16
C_{36}	–0.38	–0.39	–0.39
C_{40}	–0.37	–0.38	–0.38
C_{31}	–0.58	–0.63	–0.65
C_{33}	–0.28	–0.27	–0.27
$N(CH_3)_2$	–0.14	–0.13	–0.12

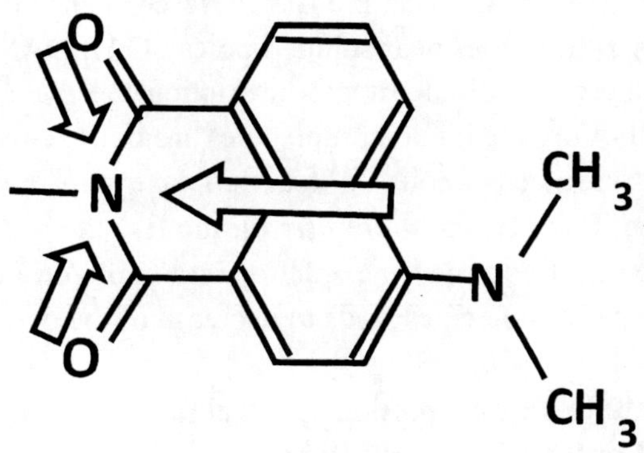

Figure 7. Dipole moments (arrows) in CAPIDAN molecule.

So, the aromatic part of the molecule can not be assumed as nonpolar and hydrophobic. Meanwhile perhaps this fragment significantly contributes to the binding energy in the case of CAPIDAN-HSA complexes. Estimation of the energy changes due to transfer of this aromatic fragment from water to nonpolar heptane has shown that this transfer requires 10 kcal/mol (i.e., is almost impossible) while the transfer from water to polar acetone gives a profit of 5 kcal/mol [92]. This profit is due to great energy of cavity formation for the fragment in water as compared with acetone. So, the aromatic fragment has poor solubility both in nonpolar media and in water but is well soluble in media of moderate polarity.

3.2. Proton Magnetic Resonance Data

CAPIDAN proton magnetic resonance spectra in water show narrow lines in 6–9 ppm area attributed to hydrogen atoms of the aromatic system, and a narrow line of 6 hydrogen atoms of the dimethylamino group at 3.0 ppm [92]. This amino group can rotate about the bond N_{35}–C_{27} but this rotation has a steric barrier: it needs ≈10 kcal/mol for the methyl group overcomes the hydrogen atom H–C_{33} [92].

4. Steady State Fluorescence

In the begging CAPIDAN-based fluorescent test was used in its steady-state version. To understand a general picture of the dye molecule environment in HSA binding sites, it has been performed a study of absorption and fluorescence spectra in a variety of solvents (Table 2). The first factor important for the spectral maxima is solvent polarity. Its magnitude can be estimated by the Lippert's function $f(\varepsilon, n)$ [93, 94]:

$$f(\varepsilon,n)=\frac{\varepsilon-1}{2\varepsilon+1}-\frac{n^2-1}{2n^2+1} \qquad (5)$$

Here n is the refractive index and ε is the dielectric constant. This function describes orientation of solvent dipolar molecules near solute dipoles . CAPIDAN molecule has dipoles (Figure 7), and therefore there is a dipole-dipole interaction the dye-solvent which leads to formation of a solvent shell of oriented dipolar molecules around the dye.

In nonpolar solvents, one needs about 70 kcal/mol to excite the CAPIDAN molecule (Table 2). The more solvent polarity, the more $f(\varepsilon, n)$, the less the excitation energy and the spectral maxima of absorption, fluorescence excitation and emission (Table 2 and Figure 8). So, solvation of the dye by polar molecules leads to decrease of the dye excitation energy.

Table 2. Absorption, excitation and emission spectral maxima of CAPIDAN in solvents of different polarity and in HSA water solutions at 25^0 C. Mean error of the maximum position is ± 1.5 nm

Solvent	Polarity $f(\varepsilon, n)$	Absorption maximum nm	Absorption energy kcal/mol	Excitation maximum nm	Emission maximum nm	Stokes' shift nm	Half-width of emission nm
Toluene	0.02	407	70.5	402	497	90	69
o-xylene	0.04	412	69.7	404	497	85	71
pentachlorhexane	0.14	397	72.3	418	502	105	70
chloroform	0.15	420	68.3	420	507	87	74
isobutanol	0.26	423	67.8	425	535	112	82
N,N-dimethyl-formamide	0.27	425	67.5	425	533	108	81
Acetone	0.28	415	69.2	425	522	107	80
Ethanol	0.29	426	67.4	428	538	112	83
Glycerol	0.26	448	64.1	449	526	77	87
water	0.32	448	64.1	448	537	89	97
human albumin	–	–	64.1	448	521	73	72

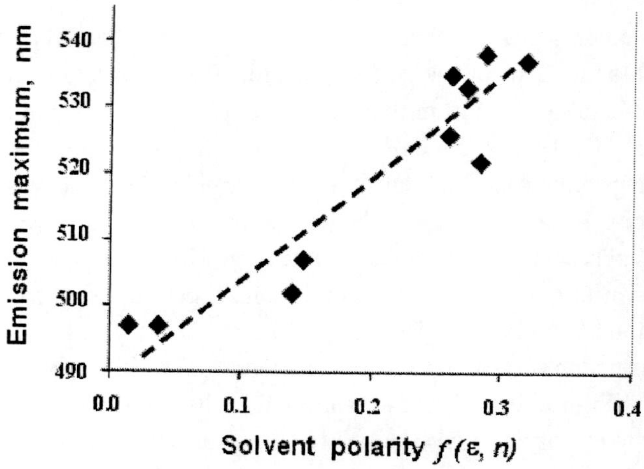

Figure 8. CAPIDAN emission spectrum in solvents of different polarity.

However this excitation energy is extremely low in the case of the HSA-bound dye: the decrease is 6 kcal/mol as compared with nonpolar media. Exactly the same energy takes place in water and glycerol. This comparison leads to suggestion that a general explanation for these three cases is the existence of hydrogen bonds solvent-solute as –O–H...O=dye. So, the dye carbonyl group or both carbonyls form hydrogen bonds in HSA binding site.

If excitation changes the dye dipole moment, it has to induce a reorganization of environmental dipoles. The reorganization needs energy, its source is the energy of the excited state. As a result, the emission maximum shifts to longer wavelength. Therefore the Stokes' shift can be more in polar media. But HSA-bound CAPIDAN has the least Stokes' shift (Table 2). Perhaps movement of polar molecules near the dye in the binding site is strongly hindered, they can not rotate in excited state life time. This hypothesis is supported by the data on rotation of the dye in HSA sites (see below, section 7.2).

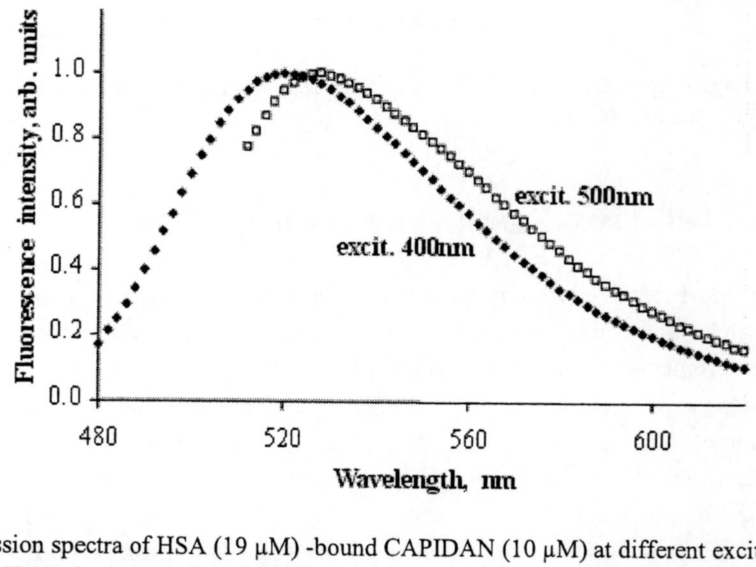

Figure 9. Emission spectra of HSA (19 μM) -bound CAPIDAN (10 μM) at different excitation wavelengths. pH 7.4; ionic strength – 0.16 M.

Structure of the polar solvation shell in polar solvents is rather variable; therefore the Stokes' shift is variable too. This phenomenon leads to widening the total emission spectrum. Indeed, the half width is about 70 nm in the case of nonpolar solvents and raises to 87 nm in polar ones (Table 2). However this dye in HSA binding sites has the same excitation maximum as in polar proton-donor solvent but an extremely narrow emission peak, 72 nm, like to nonpolar media. It can be suggested that environment of the dye in HSA is not so variable as in polar solvents, i.e., this environment in HSA has a certain structure. But nevertheless some variations take place: emission spectrum slightly moves to long wavelengths if excitation wavelength raises (Figure 9).

Degree of fluorescence polarization of HSA-bound CAPIDAN is independent of emission wavelength (Figure 10). This fact shows that the emission can be attributed to a single electronic transition at any wavelength. High anisotropy of polarized fluorescence in the excitation moment, $r(0)$, close to its theoretical maximum, 0.4, means that orientation of absorption and emission dipoles is almost similar, probably along the molecule long axis.

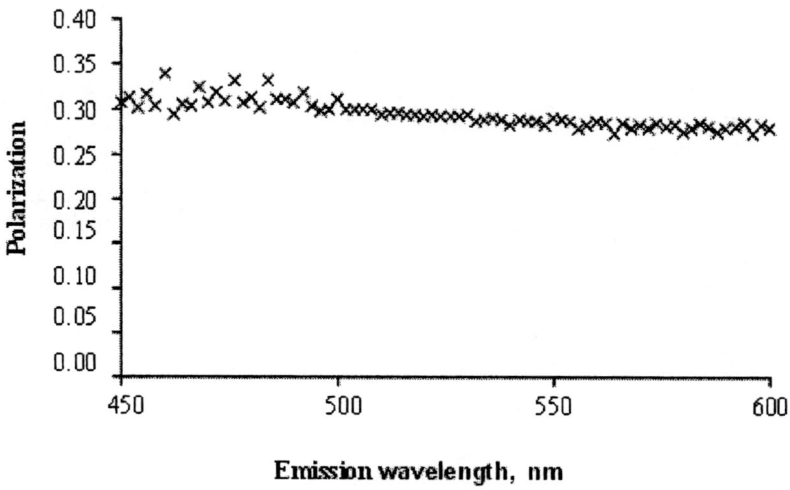

Figure 10. Anisotropy of polarized fluorescence, $r(0)$ (in the first moment after excitation), of HSA-bound CAPIDAN. Excitation 400 nm.

5. Time-Resolved Fluorescence

Steady-state fluorescence gives an interesting information on CAPIDAN environment in HSA binding sites, and many researchers use some versions of that techniques of fluorescence measurements to study of biological objects [95]. However these objects are as a rule rather complex, while steady-state fluorescence is a source of rather limited information. More over, in some cases the steady-state data give an incorrect picture. e.g., addition of phenylbuthazone, a marker of HSA drug-binding site I, to CAPIDAN-HSA complexes decreases CAPIDAN steady-state fluorescence intensity and leads to an impression that these complexes dissociate; however more detailed study shows that the complex does not dissociate but the probe environment is changed in the HSA molecule (see below section 5.4).

Much more information can be obtained in the study of fluorescence decay by the time-resolved techniques.

5.1. Simplest Cases of Fluorescence Decay

In simplest cases, there is a rather diluted solution of fluorescent molecules where environment of each molecule can be assumed as the same. C is their concentration. After illumination by a short light flash a fraction of these molecules absorbs light quanta and becomes excited. Their initial concentration is C^*_0 but it gradually decreases as a time function $C^*(t)$ due to deactivation processes. The small decrease, $-d\ C^*(t)$, of this concentration in a small time interval dt is assumed to be proportional to the concentration:

$$\frac{-dC^*(t)}{dt} = k_D C^*(t);$$

after integration: $C^*(t) = C_0^* \exp(-k_D t) = C_0^* \exp(-t/\tau_D)$ (6)

Here k_D is the rate constant of the deactivation and $\tau_D = 1/\ k_D$ is its decay time. In this simple case, three major processes can be assumed as causes of this deactivation. The first one is fluorescence emission (i.e., radiative transitions); its rate constant is k_F. The second one is intramolecular processes without fluorescence emission (i.e., nonradiative transitions); this rate constant is k_N. Finally, processes of interaction with some outer molecules could lead to deactivation, and, as a consequence, to partial fluorescence quenching; their rate constant is k_Q. As a result, k_D is the sum of these three constants:

$$k_D = k_F + k_N + k_Q \quad (7)$$

Fraction of excited molecules deactivated in time dt by fluorescence emission is $dC_F^*(t) = (k_F / k_D)\ C^*(t)\ dt$. Fluorescence intensity $F(t)$ is proportional to the derivative $dC_F^*(t)/dt$:

$$F(t) = b_C \frac{-dC_F^*(t)}{dt} = b_C \frac{k_F}{k_D} k_D C_0^* \exp(-k_D t) = b_C k_F C_0^* \exp(-k_D t) \quad (8)$$

where b_C is a coefficient dependent on parameters of the instrument used. The more two nonradiative rate constants, k_N and k_Q , the less fluorescence intensity. Preexponential factor can be named amplitude, A, of this exponential:

$$A \equiv b_C k_F C_0^*; \qquad F(t) = A \exp(-k_D t) = A \exp(-t/\tau_D) \quad (9)$$

It is interesting that both values, exponential and amplitude, depend on the radiative rate constant k_F.

This simple case can be illustrated by fluorescence decay of a fluorescent probe, 3-methoxybenzanthrone (MBA) [95, 96]. It has monoexponential fluorescence decay (straight line in Figure 11, semi logarithmic axes) with τ_D = 12.5 ns and $k_F = 4.8 \cdot 10^7$ s^{-1} in N,N-dimethylformamide.

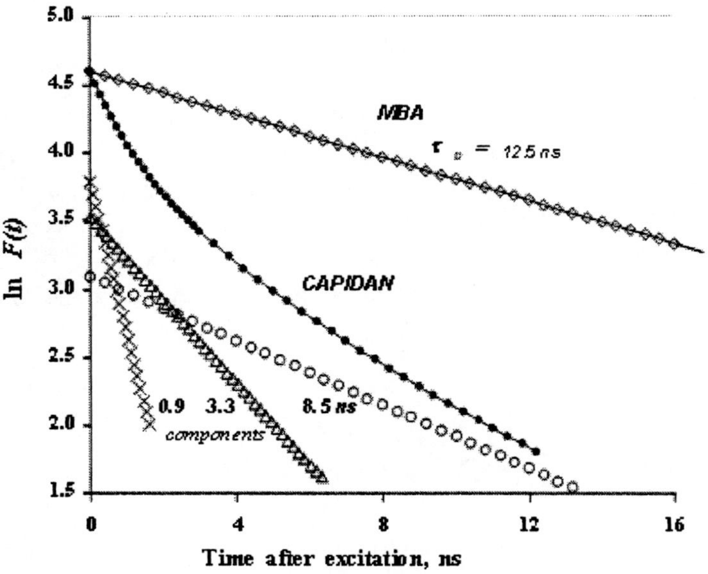

Figure 11. Fluorescence decay of MBA and HSA-bound CAPIDAN. MBA is solved in N,N-dimethylformamide. Three components of the CAPIDAN decay are shown also. Logarithmic vertical scale.

However fluorescence decay of HSA-bound CAPIDAN is not exponential (nonlinear curve in Figure 11). Obviously, either its different molecules have different environment in HSA sites or their interaction with environment in HSA molecule has a complex character.

5.2. Complicated Fluorescence Decay

Fluorescence decay of the HSA-bound CAPIDAN is described by a nonlinear curve in Figure 11, i.e., this deactivation process is not simple. This curve depends on probe and HSA concentrations, pH, ionic contents of solvents etc. Therefore the curve needs formalism for description of its shape. A convenient way to its description is a sum of exponentials:

$$F(t) = A_1 \exp(-t/\tau_1) + A_2 \exp(-t/\tau_2) + A_3 \exp(-t/\tau_3) + ... \quad (10)$$

In the case of HSA-CAPIDAN system, the sum of one or two exponentials can not describe satisfactory this decay curve while three exponentials give a good, satisfactory description [23, 92].

Table 3. Decay times τ_1, τ_2, τ_3 of albumin-bound CAPIDAN at variation of amplitude ratio A_3/A_1 due to change of ionic strength (I), phenylbuthazone (Ph) and palmitate (Pal) addition. pH 7.4. Mean error of τ_i is near ±0.1 ns. Data on fatty acid -free HSA and on sera of donors

Sample	I mol/Litre	Ph /HSA, mol/mol	Pal /HSA mol/mol	A_3 / A_1	τ_1 ns	τ_2 ns	τ_3 ns
HSA	0.02	0	0	0.85	8.9	3.4	1.0
HSA	0.19	0	0	1.80	8.8	3.1	0.8
HSA	0.19	2/1	0	6.7	8.5	2.9	0.9
HSA	0.19	0	3/1	7.7	8.7	2.7	0.8
Serum of donor 1	0.19	0	–	2.6	8.5	3.2	0.9
Serum of donor 2	0.19	0	–	5.0	8.7	3.2	0.9

However are these three exponentials an expression of some real physical processes, or it is a product of a formal mathematical procedure? If this is a mathematical artifact, then in different conditions (concentrations, pH etc.) different sets of parameters τi , Ai will take place; more over, τi would be dependent on contribution of each component, i.e., on Ai. Data in Table 3 shows that decay times τi are almost independent of Ai at very wide range of these amplitudes. So, parameters τi , Ai in the sum of Equation 10 are not a product of mathematical manipulations; they reflect actual existence of three different fluorescent species. τ1 value is close to 9 ns, τ2 is near 3 ns and τ3 is about 1 ns. Two first species are responsible for about 90-95% (40-50% each) of total fluorescence of HSA-bound CAPIDAN. This conclusion is supported by the data described below (sections 5.4 and others). Decay times are rather similar in isolated HSA and in intact donor sera (Table 3).

5.3. Estimation of Absolute Concentrations of Fluorescent Species

Equation 10 describes fluorescence decay of those molecules whose decay can be measured. Meanwhile the existence of molecules with strongly quenched fluorescence which is not measured due to their very low decay time can not be excluded.This is a problem. The second problem is the estimation of absolute concentrations of each fluorescent species. e.g., CAPIDAN fluorescence can be represented as the sum of several fluorescent species (Equation 10). However there should be a way to estimate concentrations of these species. It is clear that decay amplitudes A_i reflect these concentrations: the more the concentration, the more the amplitude (Equation 9). But amplitudes depend also on the rate constant k_F and instrument parameters.

A way to solve this problem has been proposed recently [95, 96]. In brief, a fluorescent sample studied is compared with a standard sample. There is a sample studied where concentration of the fluorescent species is C, its molar extinction at excitation wavelength is ε , fluorescence intensity in steady-state spectrum at wavelength λ is $\Phi(\lambda)$ and $S\Phi$ is integral of its steady-state fluorescence spectrum. Fluorescence decay is measured at λ. Also there is a standard solution (subscript "s") of the same chromophor with known concentration C_S and known parameters ε_S , $\Phi_S(\lambda)$, $S\Phi_S$. Their fluorescent rate constant are k_F and $(k_F)_S$, respectively. Instrument sensitivity is $b_C(\lambda)$. Approximately $\varepsilon \approx \varepsilon_S$. Then

$$\frac{A(\lambda)}{A_S(\lambda)} = \frac{b_C(\lambda)k_F C\varepsilon\left[\Phi(\lambda)/S\Phi\right]}{b_C(\lambda)(k_F)_S C_S\varepsilon_S\left[\Phi_S(\lambda)/S\Phi_S\right]} \approx \frac{k_F C\left[\Phi(\lambda)/S\Phi\right]}{(k_F)_S C_S\left[\Phi_S(\lambda)/S\Phi_S\right]} \quad (11)$$

Concentration C which needs to be measured is equal to

$$C \approx C_S \frac{A(\lambda)}{A_S(\lambda)} \frac{(k_F)_S}{k_F} \frac{\Phi_S(\lambda)/S\Phi_S}{\Phi(\lambda)/S\Phi} \quad (12)$$

If chromophor is the same in both samples, then k_F and $(k_F)_S$ could be assumed as the same, $k_F/(k_F)_S \approx 1$, and all parameters in the right part of this equation are known, then the concentration C can be calculated. However if chromophors are different, then k_F and $(k_F)_S$ are different too, and their ratio is indefinite. This case can be solved [95] using Einstein-Strickler-Berg expression for k_F [97, 98]:

$$k_F \approx const \frac{1}{(\lambda_M)^{\beta}} \quad (13)$$

where λ_M is maximum of fluorescence spectrum and $\beta \approx 3$; *const* is a combination of physical constants. This is a more general case, and Equation 12 can be modified for practical use:

$$C \approx C_S \frac{A(\lambda)}{A_S(\lambda)} \left[\frac{\lambda_M}{(\lambda_M)_S}\right]^3 \frac{\Phi_S(\lambda)/S\Phi_S}{\Phi(\lambda)/S\Phi} \quad (14)$$

where λ_M and $(\lambda_M)_S$ are maximums of fluorescence spectra of the fluorescence species studied and a standard sample, respectively. This formalism has been tested using 3-methoxybenzanthron (MBA) in organic solvents and lipid membranes and has shown satisfactory results [95]. A test has been made for the CAPIDAN case; its results are presented in Table 4. A model solution of CAPIDAN in glycerol shows that CAPIDAN fluorescence decay in glycerol is nonexponential and can be described as a sum of several exponentials (as in Equation 10). Averaged decay time

$$<\tau> = \frac{A_1\tau_1 + A_2\tau_2 + A_3\tau_3}{A_1 + A_2 + A_3} \quad (15)$$

decreases in glycerol heating. In spite of dramatic changes of the decay due to heating, the sum of amplitudes, $A_1 + A_2 + A_3$, is not changed in rather wide temperature interval. This fact could be a sign that in this temperature interval all dye molecules are fluorescent (they have decay time more than 0.2 ns). To check this hypothesis, the approach described above was used, with MBA in ethanol as a standard sample. As can see in the Table 4, the data are consistent with that hypothesis: fraction of fluorescent molecules is close to 1.0 (at T>40 ^{0}C this fluorescent fraction decreases because of appearing molecules with short decay time <0.2 ns). So, this approach can be used for estimation absolute concentrations of CAPIDAN fluorescent species.

Table 4. Fluorescence decay of CAPIDAN in glycerol at temperature variations. $<\tau>$ is a mean decay time (see Equation 15), *f*(*T*) is a fraction of fluorescent molecules having partial decay time >0.2 ns. Mean error of $<\tau>$ is ± 0.1 ns, *f*(*T*) is ± 0.13

Temperature °C	$<\tau>$, ns	*f*(*T*)
15	1.9	1.1
21	0.5	1.1
30	0.4	1.0
40	0.3	0.7

5.4. Heterogeneity of Albumin Binding Sites

As it is known, the HSA molecule is able to adsorb several aromatic dye molecules [99] but two special binding sites have the higher affinity to drugs and dyes [100] (Figure 12).

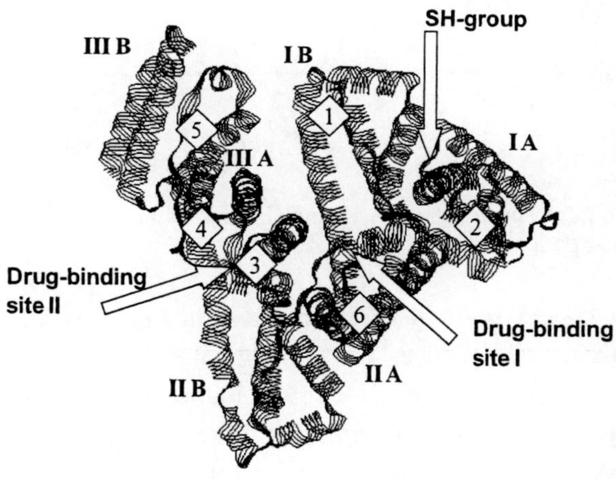

Figure 12. Localization of domains, subdomains (Roman numerals), two major drug-binding sites (arrows) and long-chain nonesterified fatty acids (rhombs with numbers) in HSA crystals. On the base of data of [101-103].

The site I is accessible for molecules of more size as compared with site II. Where does CAPIDAN localize? Why does three fluorescent species originate? Some data presented below lead to conclusion that the main place of its localization is the drug site I however interaction of the dye with this site I has different forms.

Three Fluorescent Species: Different Preexisting Environments

Steady-state fluorescence has shown that different molecules of HSA-bound CAPIDAN have different environment polarity (Figure 9). Indeed, these data are supported by time-resolved fluorescence spectra. Fluorescence decay demonstrates existence of three fluorescent species (see above). They have different decay time τ_i . When does this difference rise: during excited state or before it? Excitation spectra show that the species 1 has maximum of the spectrum near 445 nm, species 2 – near 448 nm and species 3 – near 553 nm. So, molecules responsible for these species interact with their environment differently before excitation; species 3 has the most polar solvation shell (or more strong H-bonds) than other species; the less τ_i, the more this polarity. This fact leads to suggestion that variation of the decay time τ_i may be a consequence of a different structure of polar interactions of the dye with the occupied site. May be, this polarity originates from water molecules which localize into the site I and quench dye fluorescence. This hypothesis is supported by some data described below.

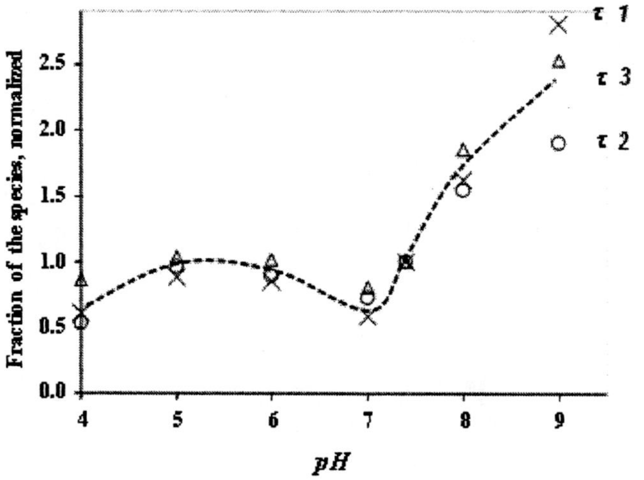

Figure 13. Fractions of the three species of HSA-bound CAPIDAN as a function of pH. Each fraction is normalized on its value at pH 7.4.

Three Fluorescent Species: Similarity of Their General Localization

As it is known, pH change induces conformational transitions in HSA molecule, so-named N–F at pH near 3-5 and N–B at pH 8–10 [99]. Fluorescence of the dye also is sensitive to these transitions (Figure 13). It is interesting that these albumin conformational transitions influence all three species similarly. Perhaps, the dye molecules responsible for these species localize in the same sites of HSA globule.

CAPIDAN Competition with a Marker of the Site I.

An anti-inflammatory drug, phenylbuthazone, is known as a marker of the drug-binding site I [99, 103]. Addition of phenylbuthazone to HSA leads to decrease of CAPIDAN steady-state fluorescence [23]. This fact can be a sign of competition of these two ligands in the site I.

The simplest scheme of this competition might be so: phenylbuthazone occupies the site, therefore the dye can not bind to the site. However time-resolved data show more interesting events. The drug influences all three fluorescence species. Three decay times are not changed (Table 3). Amplitudes of the first two species, i.e., A_1 and A_2, decrease, therefore quantity of dye molecules of this kind indeed decreases. At the same time A_3 increases. As a result, the sum A_1+A_2+A_3 which represents a total number of HSA-bound dye molecules of these species, stays approximately the same. So, phenylbuthazone changes the dye environment in the site I but does not prevent its binding to albumin. Perhaps, species 1 and 2 are converted into the species 3. Therefore decay times 9 and 3 ns are converted into 1 ns leading to fluorescence intensity decrease (Table 3).

More details of phenylbuthazone-CAPIDAN relations are described below, see Table 5.

Coulomb Interaction in the Binding Sites

All decay amplitudes of bound CAPIDAN are sensitive to ionic strength, *I*, of the solution (Table 3): the more *I*, the less amplitudes. It can mean that the negatively charged carboxyl group of the dye localizes on the boundary "site/water phase" close to some positive charges. The species have different sensitivity to the ionic strength: $A_1 > A_2 > A_3$. Amplitudes present the bound dye molecules, the binding process needs much more time than fluorescence decay time τ. So, there is a Coulomb interaction dye-albumin, but three species have different strength of the interaction. This interaction and this difference are formed long before dye excitation. In other words, the difference of environment of these three fluorescent species preexists, this is not a result of excitation.

At pH raise from 7 to 9 aminogroups of lysine residues loss their positive charge. At the same time binding of this negatively charged dye significantly increases (raise of all three amplitudes on Figure 13). So, there is no evident Coulomb interaction of the dye with lysines. Perhaps, the Coulomb interaction can be attributed to positively charged arginine residues. This hypothesis is supported by some reasons described below (section 8).

Binding Parameters of the Complex CAPIDAN-HSA

The simplest attempt to propose a formalism for the dye reversible interaction with HSA can be based on a suggestion that there exist three state of the complex CAPIDAN-site I. As a result, three fluorescent species appear. HSA concentration is C_A. The mean number of each type of sites able to form these complexes is n_1 , n_2 and n_3 per one HSA molecule, respectively. Concentration of dye molecules formed each type of these complexes is r_1 , r_2 and r_3 , respectively. Concentration of free dye is *c*. Then the simplest system of three equations for this equilibrium is

$$K_i = \frac{r_i}{c \cdot (n_i C_A - r_i)}, \qquad i = 1, 2, 3. \tag{16}$$

where K_i are the binding constants for these binding sites. Values of r_i are measured by the method of amplitude standard described above. This method permits to estimate a constant *b* in equations

$$r_1 = bA_1\,; \qquad r_2 = bA_2\,; \qquad r_3 = bA_3 \tag{17}$$

where *b* has the same value for all three species. Amplitudes A_1 , A_2 and A_3 are taken from decay kinetics (Equation 10).

Data on binding parameters in CAPIDAN-HSA complexes are presented in Table 5. The first species has the most *K*. The site I marker, phenylbuthazone, suppresses dye binding with the site decreasing both K_1 and n_1 . This marker lowers also n_2 and n_3 but strongly increases K_3 . As a result, the rough estimate of HSA affinity to the dye

$$Affinity \approx K_1 n_1 + K_2 n_2 + K_3 n_3 \tag{18}$$

is almost the same without as well as with the marker while it strongly lowers total fluorescence intensity: a rough calculated estimate for the total intensity, F_C , can be presented as

$$Fluorescence \quad F_C \approx b_F (K_1 n_1 \tau_1 + K_2 n_2 \tau_2 + K_3 n_3 \tau_3) \tag{19}$$

Presence of phenylbuthazone lowers this calculated F_C by 43 %. This is close to the experimental value 45 %. It is interesting that three partial affinities $K_i n_i$ without this marker have almost the same values (Table 5), i.e., all three species have almost the same concentrations at low ratio drug/HSA. Phenylbutazone shifts this equilibrium to the third species.

So, this simple model of HSA-CAPIDAN equilibrium allows to describe the general facts on concentrations of three bound species. More complex model takes into account a dependency of K_i on degree of site occupation by the dye [104]: K_2 and K_3 increase at occupation raises, i.e., there is a sign of positive cooperativity of the binding (however other hypotheses can be proposed too).

Table 5. Binding parameters of CAPIDAN-HSA complexes. Fatty acid-free HSA concentration is 25 μM, ionic strength 0.16 M, pH 7.4. Ratio phenylbuthazone (Ph)/HSA is 0 or 1. Mean relative error of *K* and *n* values is about ±10 %, for *Kn* values is ±13 %

Ph/HSA mol/mol	Parameter	Species		
		1 (9 ns)	2 (3 ns)	3 (1 ns)
0	*K*	45000	10000	6000
	n	0.12	0.7	1.1
	Kn	5400	6800	6600
1	*K*	27000	11000	16000
	n	0.08	0.5	0.6
	Kn	2100	5000	9600

6. CAPIDAN Solvation Inside the Binding Site

About 63 years ago, Laurence and Weber have proposed the first fluorescent probe, N-phenyl-1-amino-8-sulphonaphthalene (ANS), for studying HSA. ANS has very low fluorescence in water, but its fluorescence intensity raises many times in HSA solutions [14, 15]. CAPIDAN has the same property which is convenient for HSA study: water quenches its fluorescence, so in water the dye has a low decay time $\tau_W < 0.2$ ns. Binding to HSA hides the dye from water, and decay time raises. At the same time a fraction of the HSA-bound dye is partially quenched: its decay times, $\tau_2 \approx 3$ ns and $\tau_3 \approx 1$ ns , are significantly lower than the high value of $\tau_1 \approx 9$ ns. What is the quencher? What is the quenching mechanism in water and in HSA?

Both steady-state and time-resolved techniques show (see above) that all three species of bound dye molecules have very polar environment and/or H-bonds with environment, as in water. It is probable, some water molecules penetrate the binding site and solvate the dye molecule. However a simple solvation by dipole-dipole interaction is not able to shift absorption and excitation spectra to the red side so strongly (Table 2). Therefore this shift could be a result of H-bonds between water molecules and O=C groups of the aromatic chromophor, i.e., HO–H...O=C–dye.

Can the polar medium and/or H-bonds be a major cause of the partial fluorescence quenching of the bound dye? High environment polarity shifts electron levels in chromophors having dipole moments, distances between levels are changed leading to change probability of radiationless processes, i.e., influencing efficiency of quenching (see e.g., [105, 106]). However the course of studying ANS quenching by water has shown that polarity was not able to explain this quenching. A hypothesis has been proposed for the ANS case [107]: a cluster of 3-4 water molecules is formed near ANS amino group, and electron of excited ANS* is transferred to the water cluster. Obviously, this transfer takes place due to H–bonds ANS*–N–H...water. As a result, fluorescence is quenched. H-bond is a bridge for this charge transfer.

Last years, direct data on fluorescence quenching by electron transfer between chromophors and water (or other molecules forming H-bonds) appear (e.g., see [105-110]). In these cases, formation and existence of H-bond are assumed as the cause of the quenching. However, on the other hand, recently some facts were discussed about a probable role of H-bond breaking in the quenching [111]. Indeed, in water as well as in glycerol CAPIDAN has a very "red" absorption spectrum (Table 2). This fact can be a sign of existence H-bonds of the dye carbonyl group(s) =O with OH-groups of these solvents. Excitation spectra have the same "red" peak position as absorption spectra. Coincidence of these excitation and emission peaks could be assumed as a sign of (i) participation of all dye molecules in H-bonds with these solvents, and (ii) participation of all dye molecules in fluorescence emission. These conclusions are supported by the direct data on the fraction of emitting molecules (Table 4): in glycerol all 100 per cent of CAPIDAN molecules participate in the emission.

So, existence of H-bonds solute-solvent in this case does not prevent fluorescence emission. Glycerol heating decreases significantly decay time of this emission (Table 4), i.e., quenches fluorescence; however number of fluorescent molecules stays the same. Heating decreases glycerol viscosity, therefore it can be expected raise of H-bonds breaking. So, this breaking could be assumed as a cause of fluorescence intensity decrease. This hypothesis is supported by some reasons on intramolecular charge transfer at H-bond disappearance [111].

Returning to HSA, this hypothesis can be applied to HSA-bound CAPIDAN. It has the same maximum of excitation spectrum as in water and glycerol (Table 2); unlikely this is an accidental coincidence. Probably, the dye forms H-bond(s) with H-donating groups of the binding site, may be with water which is entrapped by the site. Sites with immobile molecular dynamics, with long-lived H-bonds, give the first dye species with high decay time about 9 ns. Sites with rather mobile interior and short-lived H-bonds give the second and third species with the time of 3 and 1 ns. This conclusion is supported by a direct data on HSA-bound CAPIDAN mobility (see section 7.2): the more the mobility, the less the decay time. In water phase the H-bonds dye=O...H–OH have very short lifetime, therefore fluorescence decay time and intensity are very low (τ < 0.01 ns).

7. Dye Rotational Mobility in Heterogeneous Albumin Binding Sites

Molecular motion in the binding sites has been tested by proton magnetic resonance (^{1}H NMR) and by time-resolved polarized fluorescence of the bound dye.

7.1. Proton Magnetic Resonance Data

Free CAPIDAN in water has a narrow ^{1}H NMR line 3.0 ppm due to 6 protons of dimethylamino group and narrow signals of protons of aromatic rings in the area 7.0–8.4 ppm (Table 6 and Figure 14) [92]. Height of all these lines falls after HSA addition. So, molecular motion around the dye molecule significantly restricted due to its binding with HSA. The amino line 3.0 ppm is well seen on the HSA background signals and therefore is especially sensitive to the binding [92]. Probably, its rotation restricts in HSA molecule; i.e., the group is inside the binding site.

Table 6. CAPIDAN hydrogen atoms responsible for ^{1}H NMR lines. Deuterium water pD 7.4. Atom numbers are shown in Figure 14. On the basis of [92]

Hydrogen atom	37-39, 41-43	34	32	30	26	24	8; 9	7; 10
Chemical shift, ppm	3.0	8.1	7.5	8.4	7.0	8.3	7.3	8.0
singlet/doublet/triplet	s	d	t	d	d	d	d	d

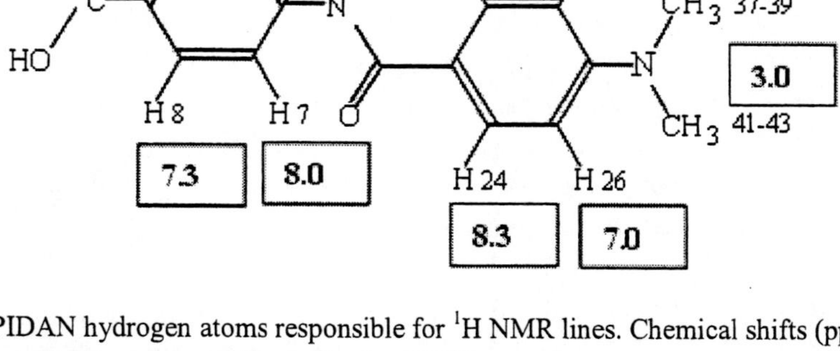

Figure 14. CAPIDAN hydrogen atoms responsible for ^{1}H NMR lines. Chemical shifts (ppm) in deuterium water pD 7.4 are framed. On the basis of [92].

A rather strong Coulomb interaction of the dye COO^- group with HSA charges (see sections 5 and 8) at physiological ionic strength can signify that the distance between them is rather low, not more than several Angstrom. NMR data shows that the dye amino group on the opposite side of this molecule is hidden of the free water phase too. So, the entire dye molecule is hidden in the HSA binding site (except of COO^- in the boundary site/water phase).

7.2. Time-Resolved Polarized Fluorescence Data

This approach permits to estimate rotational mobility of each CAPIDAN species separately in HSA binding sites. If a dye solution is irradiated by vertically (index V) polarized light, then its emission is polarized too. At pulse excitation, emission decay F(t) can be described as a sum of vertically (index VV) and horizontally (index VH) polarized components:

$$F(t) \;=\; F_{VV}(t) \;+2F_{VH}(t) \tag{20}$$

Anisotropy of polarized fluorescence, $r(t)$, decays exponentially in homogeneous solutions [94, 112, 113]:

$$r(t) \;=\; \frac{F_{VV}(t) - F_{VH}(t)}{F_{VV}(t) + 2F_{VH}(t)} = r(0)\exp(-t/\psi) \tag{21}$$

where $r(0)$ is anisotropy at $t = 0$ and ψ is an anisotropy decay time constant (or rotational correlation time). The fluorescence polarized components decay as

$$F_{VV}(t) = A_V \exp(-t/\tau) \;+ A_V 2r(0)\exp(-t/\gamma)$$
$$F_{VH}(t) = A_V \exp(-t/\tau) \;- A_V r(0)\exp(-t/\gamma)$$

where $\gamma = \psi\tau/(\psi+\tau)$ (22)

So, these polarized components decay nonexponentially, each of them is a sum of two exponentials. In heterogeneous cases a set of m decay times τ_i and ψ_i exists. The expressions 22 can be generalized [112, 113] to

$$F_{VV}(t) = \sum_{i=1}^{m} [A_{Vi}\cdot\exp(-t/\tau_i) \;+ A_{Vi}2r_i(0)\cdot\exp(-t/\gamma_i)]$$
$$F_{VH}(t) = \sum_{i=1}^{m} [A_{Vi}\cdot\exp(-t/\tau_i) \;- A_{V_i}r_i(0)\cdot\exp(-t/\gamma_i)\;] \tag{23}$$

e.g., in the case of a minimal heterogeneity ($m = 2$) each polarized component is a sum of 4 components. Separation of them is not a simple problem [112, 113]. In the case of HSA-bound CAPIDAN $m = 3$, each decay component contains 6 exponentials. Resolution it into 6 components gives a mathematically unstable solution; in other words, a general solution of this problem in analysis of a real experimental data is absent. However recently has been shown [114], that if $\psi >> \tau$, then each of $F_{VV}(t)$, $F_{VH}(t)$ can be approximated by a single exponential instead of two. Indeed, at $\psi >> \tau$ parameter $\gamma = \psi\tau/(\psi+\tau) \approx \tau(1+ \tau/\psi)$ is close to τ, and two exponentials in Equations 22 (as well as in 23) with almost similar decay times can be assumed as a single one with an intermediate decay time τ_{Vi} for $F_{VV}(t)$ or τ_{Hi} for $F_{VH}(t)$. e.g., in the case of $\psi = 5\tau$ the error of that approximation is about 1 per cent. So, at $m = 3$:

$$\begin{aligned} F_{VV}(t) &\approx \sum_{i=1}^{3} A_i \cdot [1 + 2r_i(0)] \cdot \exp(-t / \tau_{Vi}) \\ F_{VH}(t) &\approx \sum_{i=1}^{3} A_i \cdot [1 - r_i(0)] \cdot \exp(-t / \tau_{Hi}) \end{aligned} \quad (24)$$

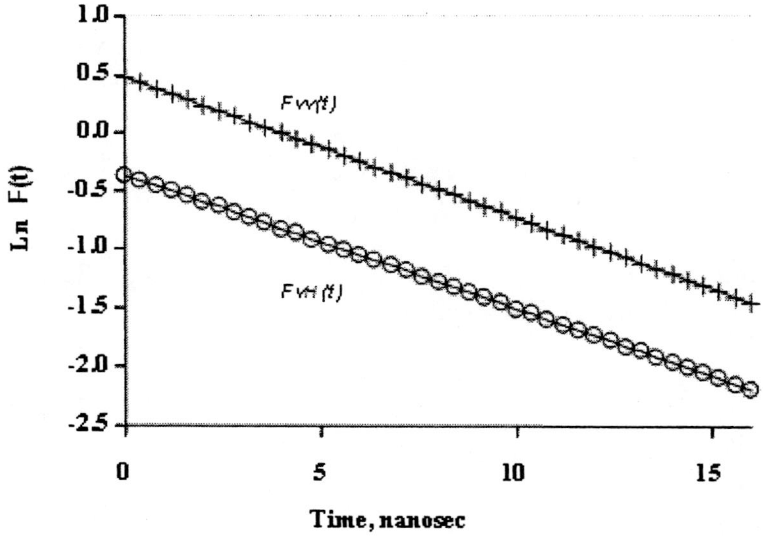

Figure 15. Decay of polarized components of the first species of HSA-bound CAPIDAN. Fatty acid free HSA, pH 7.4. $\psi/\tau = 8.5$.

So, decay of $F_{VV}(t)$ and $F_{VH}(t)$ should be measured, each of them should approximated by 3 exponentials, this procedure gives 3 pairs of τ_{Vi} and τ_{Hi} . Difference between them, i.e., ($\tau_{Vi} - \tau_{Hi}$), allows to calculate times of rotational relaxation ψ_i for each species of the bound dye. This parameter expresses time necessary for deviation of the dye long axis (Figure 16) from its initial direction at excitation moment. If these deviations have an isotropic character, the coefficient of rotational diffusion is equal to

$$D_{Ri} = k_B T / (8\pi b^3 \eta_i) \quad (25)$$

where b is half of the dye size in the direction along the long axis, η_i is environment viscosity for this species, k_B is Boltzmann constant and T is temperature.

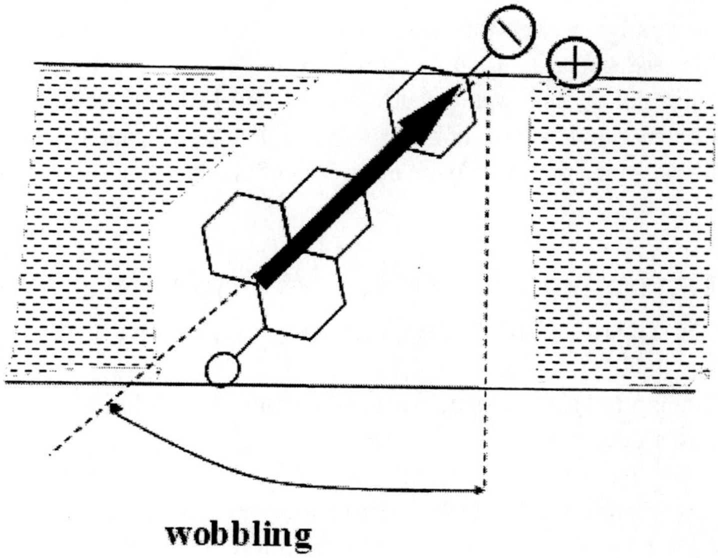

Figure 16. Hypothetical scheme of CAPIDAN wobbling in HSA binding site.

In the case of CAPIDAN bound to fatty acid free HSA the species 1 has $r(0)$ close to 0.30 and $(\tau_{Vi} - \tau_{Hi}) = (8.23 - 8.79) = -0.56$ ns; decay of its polarized components can see in Figure 15. Data of the bound dye species mobility are presented in Table 7.

As has been proposed above, one (carboxylic) end of the dye molecule is tightly bound with a positive charge(s) of the binding site (see sections 5 and 8). But a rotational freedom exists. The first species has ψ equal or more than an estimate for rotation of HSA globule (40–80 ns). So, this species is really immobile in its binding site. Probably, its motion is strongly limited. The second and especially third species move rather free.

Table 7. Rotational mobility, ψ_i (ns), of HSA-bound CAPIDAN in isolated HSA (fatty acid free) without or with palmitate addition and in serum of a healthy donor

Sample	Species		
	1 "9 ns"	2 "3 ns"	3 "1 ns"
HSA fatty-acid free	72 ± 7	41 ± 8	9.6 ± 1.3
HSA + palmitate = 1: 0.8 (mol/mol)	45 ± 5	15 ± 5	6 ± 2
Serum	49 ± 4	21 ± 3	8 ± 2

As has been proposed above, one (carboxylic) end of the dye molecule is tightly bound with a positive charge(s) of the binding site (see sections 5 and 8). But a rotational freedom exists. Therefore it can be suggested that other end of the dye molecule is relatively free for movement. Then this movement is probably like to wobbling of the amino-end (Figure 16). That wobbling of the second and third species is a probable cause of fluorescence depolarization. This free motion could facilitate fluorescence quenching by water and, as a consequence, decrease of the dye decay times τ_2 and τ_3 up to 3 and 1 ns, respectively. In the sites for the first species so fast motion is absent, therefore τ_1 has its maximal value. Wobbling of other aromatic ligands bound to HSA has been supposed earlier (e.g., [115]).

I It is interesting that this motion in serum is faster than in isolated, fatty acid free HSA Table 7). Perhaps, this is an effect of fatty acids: e.g., palmitate addition to fatty acid free HSA leads to decrease ψ (i.e., raise of motion) up to level of HSA in serum (Table 7). Palmitate addition in this sample was equal to a mean level of fatty acids in natural sera of healthy donors. This motion is almost nonsensitive to pH changes from 4 to 9; so, N-F and N-B conformational transitions do not influence freedom of the dye rotational motion in HSA.

8. Chloride/Nitrate Substitution for Study of Drug-Protein Coulomb Interaction

CAPIDAN binding to HSA depends on ionic strength, I, of solution. Initial fluorescence intensity, F_0 , at low I decreases at KCl addition, i.e., at I raise. Effect of this F_0 decrease to F can be expressed as $(F_0 - F) / F_0$. Ionic strength is proportional to square root of KCl concentration in accordance with the Debye-Huckel theory. Indeed, in the case of KCl there is a dependency of this effect on $I^{0.5}$ rather close to straight line (Figure 17). There exist similar examples of other dyes which interaction with HSA is sensitive to ionic strength and is described as a linear function of $I^{0.5}$ (e.g., [115]) though the binding dye-protein should be described by a much more complicated function.

All amplitudes of the three CAPIDAN species lower due to I raise, i.e., concentrations of bound dye of all species fall (Table 9) [116, 117]. The most sensitive to ionic strength is the first species: the raise I from 0.02 to 0.52 M causes 3-fold decrease of r_1 (Table 9 and [116]), and from 0.002 M to 0.52 M – by 5.1-fold [92]. If it is assumed that the ionic strength prevents Coulomb interaction of the dye negative charge with HSA positive charges, then the free energy ΔG of the interaction can be estimated [92]. This free energy at $I = 0.002$ M is equal to $\Delta G_{0.002}$ and at 0.52 M – to $\Delta G_{0.52}$. It can be assumed that $|\Delta G_{0.002}| >> |\Delta G_{0.52}|$ in accordance with Debye-Huckel theory. So, a transition from $I = 0.52$ M to $I = 0.002$ M for the first species gives a change of the free energy of dye-site interaction:

$$\Delta G_{.0.002} \approx \Delta G_{0.002} - \Delta G_{0.52} \approx -RT \cdot \ln(5.1) = -1\, kcal / mol \qquad (26)$$

On the basis of the data presented in Figure 17, it could be estimated that the Coulomb energy of the first species lowers to –0.4 kcal/mol at physiological ionic strength [92]. The similar interaction takes place in the cases of the second and third species but their energy is lower; may be, in these cases distances between charges dye/albumin are larger (this hypothesis is supported by facts of more rotational mobility, see section 7.2 and Figure 16). These estimations are rather approximative.

On the other hand, if chloride anion is replaced by nitrate anion, then fluorescence decrease is nonlinear function of $I^{0.5}$ and rather stronger than in the case of chloride (Figure 17). So, nitrate causes by (i) raising ionic strength but (ii) has an additional, specific ability to decrease of HSA-CAPIDAN fluorescence. Either nitrate decreases its decay time τ, or it suppresses dye binding. However τ of all three species stays almost unchanged at nitrate addition. Titration of HSA by the dye (Figure 18) has shown that the intercept on the vertical axis is the same for both anions. So, number of binding sites is unchanged at replacing chloride by nitrate. On the other hand, the intercept on the horizontal axis is decreased under this replacement. This intercept is an estimation for the binding constant. So, nitrate suppresses the binding constant K_1, e.g., by 2.9 times at this partial replacement of chloride by nitrate. K_2 is 1.8-fold decreased under the same conditions, K_3 is decreased too.

Thus, a specific effect of nitrate as compared with chloride is suppressing the dye binding to HSA by decreasing the binding constant. What is the molecular basis of that effect? It can be assumed that nitrate suppresses the Coulomb interaction of the dye negative charge and positive charges on the interface of the site I and water phase, i.e., HSA lysine and arginine residues (Figure 19). Either nitrate anion specifically binds to lysine neutralizing its positive charge or it neutralizes the arginine charge. Conductometric experiments on charge interaction in electrolyte solutions have shown [116, 117] that the nitrate anion has significantly more affinity to guanidine cationic group of arginine than chloride anion.

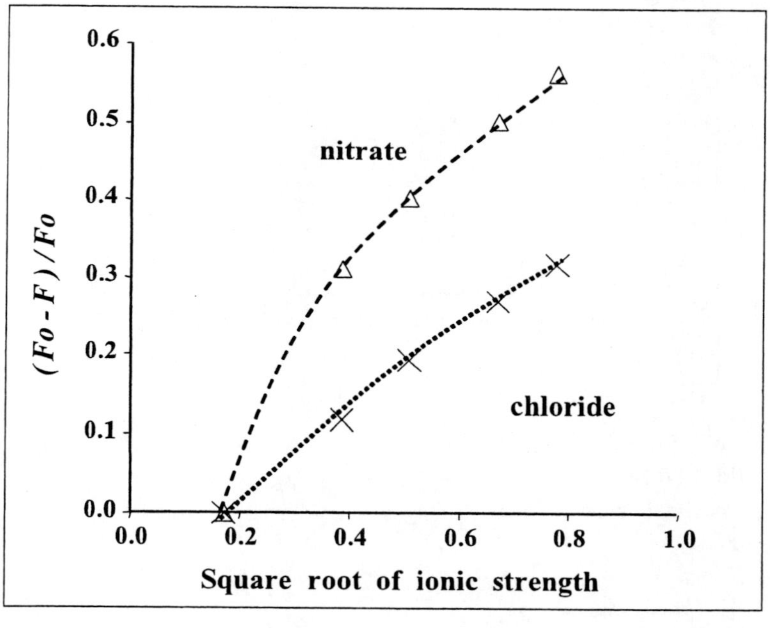

Figure 17. Decrease of CAPIDAN fluorescence intensity due to KCl or KNO_3 addition to solution of CAPIDAN and HSA in 0.01 M phosphate pH 7.4. HSA was fatty acid free.

So, it can be suggested that the negatively charged group –COO^- of CAPIDAN interacts with an arginine guanidine group on the interface of the site I/water phase. This arginine could be either Arg 218, Arg 222 or Arg 257 (Figure 19). Binding of another anionic drug, warfarin, to the same site significantly enhances after Lys199 replacement by an uncharged residue but falls threefold after Arg218 replacement [102, 118]. These data support the suggestion that CAPIDAN also interacts with arginines of the site I, may be with Arg 218, and nitrate anions hinder this interaction.

That technique of nitrate/chloride replacement can be used for detecting the role of Coulomb interaction arginine-ligand in ligand binding by proteins.

Table 9. Influence of ionic strength I on CAPIDAN binding to HSA. r_1, r_2 and r_3 are concentrations of the dye species bound to HSA; $r_i / r_{i(0.02)}$ – these concentrations are normalized to their values at I = 0.02 M. Fatty acid free HSA, pH 7.4. Data from [116]

Parameter	Ionic strength I, mol/Litre		
	0.02	*0.16*	*0.52*
$r_1 / r_{1(0.02)}$	1.00	0.52	0.32
$r_2 / r_{2(0.02)}$	1.00	0.63	0.44
$r_3 / r_{3(0.02)}$	1.00	0.80	0.62

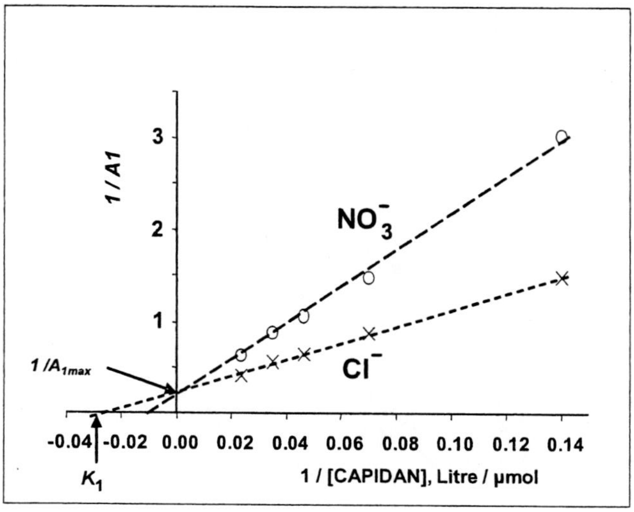

Figure 18. Titration of HSA by CAPIDAN in the presence of chloride 0.63 mol/litre and after partial equimolar substitution of 0.38 mol/litre chloride by nitrate. A_1 is amplitude of the CAPIDAN first species, A_{1max} is its maximal value. K_1 is the binding constant. Fatty acid free HSA, pH 7.4. On the basis of the data from [117].

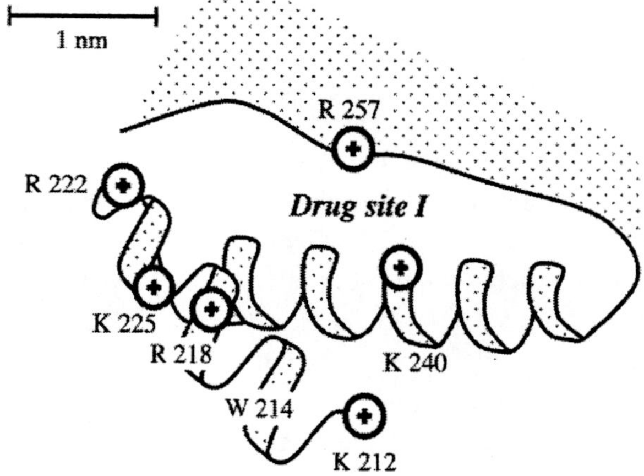

Figure 19. Schematic representation of positive charges on the surface of the Site I in HSA crystals. On the basis of the crystallographic data of [102, 119, 120]. K is lysine, R is arginine, W is tryptophan.

9. THERMODYNAMICS OF CAPIDAN-ALBUMIN INTERACTION

The fact of CAPIDAN binding to HSA signifies in the formal language of physics that free energy of this solution decreases as a result of the complex formation. This decrease is ΔG. The simplest formalism for HSA-dye interaction could be as follows. Two processes take place: (i) association of the dye with the HSA binding site and (ii) dissociation of their reversible complex. Rate constants of these processes are k_{Ass} and k_{Dis} (Figure 20). To form such complex, it needs to overcome a barrier ΔG_A, its dissociation needs ΔG_D (Figure 20). If the difference $\Delta G = (\Delta G_A - \Delta G_D) < 0$, then the complex is thermodynamically probable.

The binding constant (see Equation 16) $K_B = k_{Ass} / k_{Dis}$, and the rate constants can be assumed as exponential functions of ΔG_A, ΔG_D . The free energy includes two components: enthalpy ΔH and entropy ΔS. So,

$$K_B = \frac{k_{Ass}}{k_{Dis}} = \frac{k_{A0}\exp(-\Delta G_A / RT)}{k_{D0}\exp(-\Delta G_D / RT)} = \frac{k_{A0}\exp(-(\Delta H_A - T\Delta S_A)/RT)}{k_{D0}\exp(-(\Delta H_D - T\Delta S_D)/RT)} = \\ = \frac{k_{A0}}{k_{D0}}\exp(-\frac{\Delta H_B - T\Delta S_B}{RT}) = \frac{k_{A0}}{k_{D0}}\exp(-\frac{\Delta G_B}{RT}) \qquad (27)$$

Then

$$\ln K_B = [(\ln \frac{k_{A0}}{k_{D0}}) + \frac{\Delta S_B}{R}] \; - \frac{\Delta H_B}{RT} \qquad (28)$$

This equation shows that measuring K_B at different temperatures T can help calculate enthalpy of the binding, i.e., ΔH_B . However estimation of entropy, ΔS_B, is not so simple because the temperature-independent part of the Equation 28 includes a ratio k_{A0}/k_{D0} besides of entropy, and there is no a simple way to estimate this ratio.

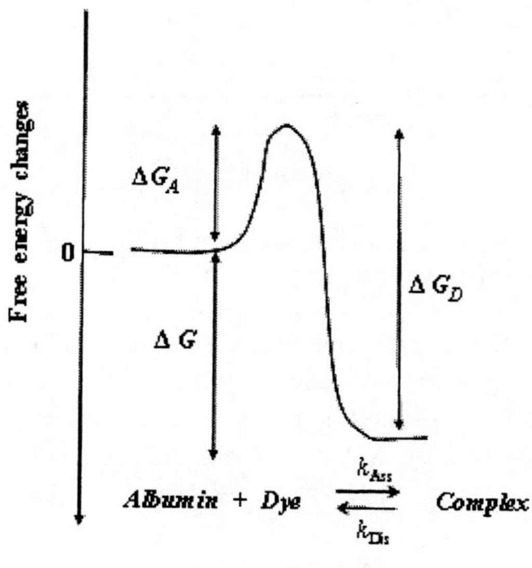

Figure 20. Scheme of energetic barriers at HSA-CAPIDAN complex formation.

Table 10. Enthalpy, ΔH_B, of binding CAPIDAN to HSA. Data of a formal calculation on the basis of Equation 28 for the temperature interval from 15 to 45 ^{O}C. Fatty acid free HSA. pH 7.4. Mean error ± 0.2 kcal/mol

Species	Ionic Strength, M	ΔH_B, kcal/mol
1	0.02	0.15
1	0.16	−1.3
1	0.6	−3.6
2	0.16	−1.4
3	0.16	−0.7

In the case of CAPIDAN-HSA interaction, both parameters of fluorescence decay of each species, A_i and τ_i, are dependent on temperature. The first of them, A_i, reflects the concentration of bound dye, while the second, τ_i, depends on quenching processes in the binding site. Formally calculated values of enthalpy, ΔH_B, of CAPIDAN binding to HSA are shown in Table 10.

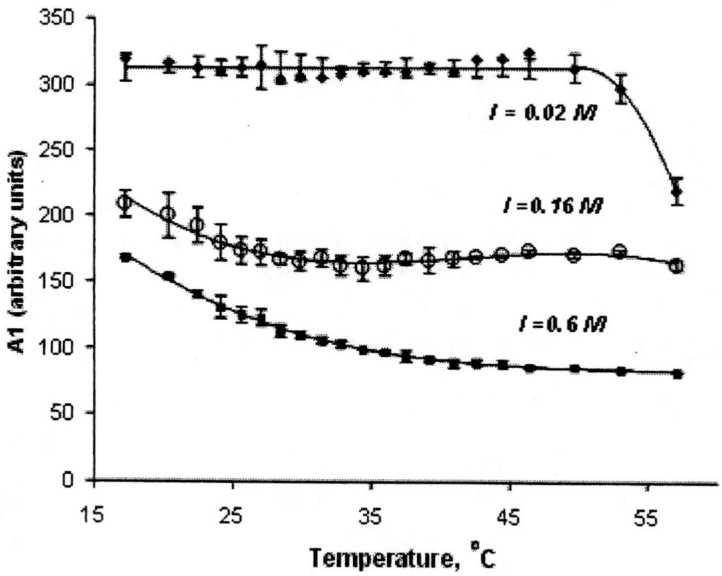

Figure 21. Dependency of the amplitude A_1 of the CAPIDAN first species on temperature T at three ionic strengths I. Fatty acid free HSA. pH 7.4.

Dependency of A_1 on T at different ionic strength is shown in Figure 21. As can be seen, the upper curve (low ionic strength) has a complex form. At 15-48 °C there are not significant A_1 changes. So, here the binding of the first species is independent of T. This fact formally could mean that the binding enthalpy, ΔH_B, is close to zero (Table 10). However the binding exists, therefore

$$\Delta G_B = \Delta H_B - T\,\Delta S_B < 0; \text{ then } \Delta H_B \approx 0; \Delta S_B > 0 \quad (29)$$

i.e., the driving force of the binding could be a raise of entropy: $T\Delta S_B > 0$. In this temperature interval raising I leads to increasing of the slope of the curves in Figure 21, i.e., binding at physiological I and especially at high I is decreased at heating. So, formally the value of enthalpy ΔH_B grows at ionic strength increasing.

These results seem paradoxical. All experiments show existence of Coulomb interaction the dye-HSA which is suppressed by ionic strength (see above, section 8). This tendency is seen in Figure 21: amplitudes decrease at ionic strength increase. At the same time presence of the Coulomb force should be expressed by rather high enthalpy which should decrease up to ≈ 0 at high I. However results of Figure 21, as it seems, are opposite.

To try solving this paradox, the next hypothesis can be proposed. Enthalpy ΔH_B can be calculated using Equation 27 if its parameters, i.e., k_{Ass}/k_{Dis} , ΔH_B , ΔS_B are temperature-independent constants. However the binding site I has a rather complex structure, and these parameters could be interdependent or/and temperature-dependent. The first explanation of the horizontality of the 15-48 ^{O}C interval on the upper curve, i.e., $\Delta H_B = 0$, seems unlikely (see above). The second formal explanation is an enthalpy raise at heating:

$$\Delta H_B \approx -\mathrm{h}T \text{ were h>0 is a constant.} \qquad (30)$$

The interval of 48-55 ^{O}C of the upper curve supports in general this hypothesis: the more T, the more slope of the curve, i.e., formally the more ΔH_B. The third case may be a raise of the ratio k_{Ass}/k_{Dis} or entropy ΔS_B at heating.

Such clear deviations from the simplest model (Equation 27) occur only at low ionic strength. May be, these anomalies are due to mutual repulsion of positive charges in their cluster on the interface Site I–Water phase (Figure 19). This repulsion enhances at low ionic strength and leads to swelling of the cluster: charges move apart, the space is filled up by water. The anomalies at low ionic strength are an effect of the dye negative charge and temperature on this swelling of the positively charged cluster.

10. FATTY ACID INFLUENCE ON THE DRUG-BINDING SITES

Transfer of long-chain nonesterified fatty acids (FA) by HSA is one of the main HSA functions. FA concentration in plasma is a risk-factor for sudden cardiac death [121, 122]. Electron paramagnetic resonance signals of spin-labeled FA were significantly changed at cancer and other diseases [123, 124]. FA content of plasma depends on motor activity of rats at stress [125] etc. So, FA level is tightly associated with physiological, emotional and pathological status.

Plasma FA are mainly bound to HSA. Mean ratio FA/HSA in blood is close to 0.7–1 mol/mol [36] but HSA molecule is able to bind even 10-12 FA molecules. One site for FA binding has the binding constant about 10^7 M^{-1}, two sites have 10^6 M^{-1} and 5 sites have near $10^3 - 10^4$ M^{-1} [36]. Localization of sites for the first six FA molecules (i.e., sites of the most binding constant) in HSA crystals is shown in Figure 12 [126, 127].

FA influence binding of fluorescent [128-131] and spin [132] probes and other ligands to HSA. Binding of some ligands is suppressed while of others is enhanced by FA addition. CAPIDAN fluorescence enhances due to addition of 1-3 FA molecules per one HSA molecule [23, 133]. Enhancement of the total fluorescence may be a result of decay time or/and amplitude changes. In the case of FA addition to HSA all three decay times of the three decay CAPIDAN species are slightly lower while total fluorescence intensity increases significantly due to raise of amplitudes; i.e., FA increase binding of the dye to HSA. However sensitivity of the three main species of the probe to FA is different: the first one has a minimal sensitivity and the third one has maximal (Figure 22).

What is a mechanism of this binding raise? FA localization in HSA crystals is shown in Figure 12 (see section 5). Negatively charged carboxyl of the FA molecule number 2 localizes near positively charged Arg 257 in the site I [119]. It seems that in water solutions the negatively charged FA carboxyl should be attracted to the positively charged cluster on the surface of the site I (see Figure 19 in section 8). As a result, the binding constant for this negatively charged dye is expected to be decreased after FA appearing. But it raises. More over, the first species is the most sensitive to the charge of the cluster (see section 8), however it is almost insensitive to FA (Figure 22). The third species, in contrast, is almost insensitive to ionic strength but its population increases by 3 times at addition of 3 FA molecules. FA effects on the dye binding are almost independent of ionic strength of solution; this is an indirect evidence of absence Coulomb interaction of FA and the dye in HSA.

So, the FA negatively charged carboxyl unlikely localizes in the positive cluster of the site I in water solutions. There are no evidences of competition of the dye and the first three FA molecules. Quite the contrary, these FA molecules enhance the dye binding to HSA. Physical chemical causes of this situation need to be clarified.

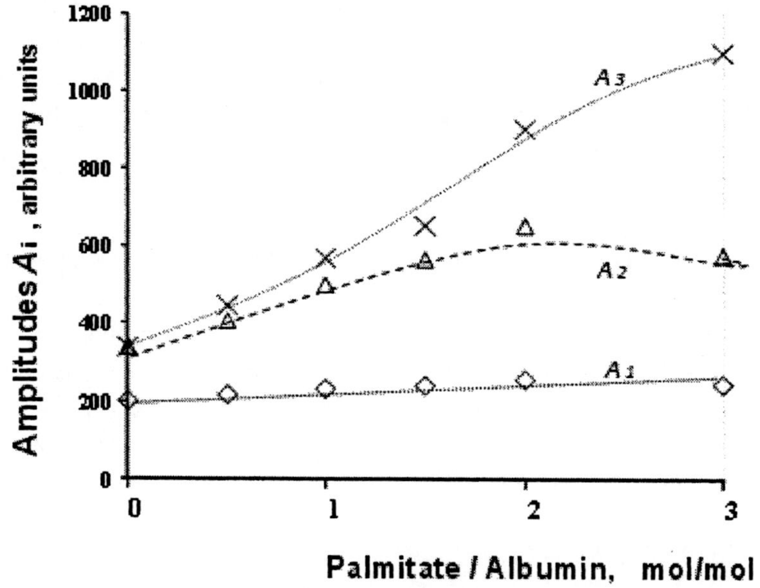

Figure 22. Influence of palmitate addition to HSA on amplitudes of three species of CAPIDAN fluorescence decay. Initial HSA was fatty-acid free. pH 7.4, ionic strength 0.16 M.

CONCLUSION

A lot of aromatic substances, both fluorescent and nonfluorescent, have been used to study HSA drug binding sites. CAPIDAN is one of those substances. After its addition to diluted plasma or serum the main part of its fluorescence, 90–95 per cent, originates from its complexes with HSA, in contrast with many other aromatic substances. This property leads to study of HSA binding sites in their intact state without any extraction and fractionation procedures. Therefore this dye has been used to detect changes of the binding sites which accompany some pathological processes. In some cases this information is useful for clinical diagnostics exceeding routine clinical analyses. The sensitivity to site changes is raised by using time-resolved measurements of fluorescence decay and three new methods of the data analysis: (1) estimation of absolute values of concentration of fluorescent species in heterogeneous biological objects; (2) measurement of dye rotational mobility in heterogeneous albumin binding sites; and (3) a method of chloride/nitrate substitution to detect interaction of a drug negative charge with arginine positive charges in the albumin binding sites.

A part of disease-accompanied changes of HSA molecule can be attributed to variations of quantity and content of HSA-bound nonesterified fatty acids; however these variations can not explain in full measure the structural changes of the binding sites detected by CAPIDAN.

REFERENCES

[1] Grollman, A. (1925). The combination of phenol red and proteins. *J. Biol. Chem.,* 64, 141-160.

[2] Goldstein, A. (1949). The interactions of drugs and plasma proteins. *Pharmacol Rev*, 1, 102-165.

[3] Lindup, W. E. (1975). Drug-albumin binding. *Biochem. Soc. Transact.*, 3, 635-640.

[4] Kosh-Weser, J. & Sellers, E. (1976). Binding of drugs to serum albumin. *New England J. Med.*, 294, 311-316.

[5] Richardson, A., Miller, I., Schumacher, C., Jamber, W., Pancy, F. & Lapedes, D. (1946). Physiological disposition of penicillin G and K in dogs. *Proc. Soc. Exper. Biol. Med.*, 63, 514-519.

[6] Klotz, I. M. (1946). Spectrophotometric investigations of the interactions of proteins with organic anions. *J. Amer. Chem. Soc.*, 68, 2299-2304.

[7] Klotz, I. M., Burkhard, R. & Urquhart, J. (1952). Structural specificities in the interaction of some organic ions with serum albumin. *J. Amer. Chem. Soc.,* 74, 202-208.

[8] Klotz, I. M. & Walker, F. M. (1947). Binding of organic ions by proteins - charge and pH effects. *J. Amer. Chem. Soc.*, 69, 1609-1612.

[9] Frederichsen, P. & Kierulf, P. (1979). A more accurate dye-binding method for the routine determination of serum albumin. *Clin. Chem.*, 25, 1180.

[10] Doumas, B. T., Watson, W. A. & Biggs, H. G. (1971). Albumin standards and the measurement of serum albumin with bromcresol green. *Clin. Chim. Acta*, 31, 87-96.

[11] Csögör, S. I. (1972). Functia de transport a serumalbuminei (Bucuresti, Republic Socialiste Romania: Editura Academiei.

[12] Melnik, I. A., Baranovsky, P. V. & Nesterenko, L. I. (1985). [A new method for estimation of serum albumin transport function]. *Lab. Delo,* 4, 202-204 (Russian).

[13] Salomatin, V. V. & Lifshitz, R. I. (1987). [Binding capacity of serum albumin at thermal burn]s. *Vopr. Med. Khim.*, 33, 73-77 (Russian).

[14] Laurence, D. J. R. (1952). Adsorbtion of dyes on bovine serum albumin by the method of polarization of fluorescence. *Biochem. J.,* 51, 168-180.

[15] Weber, G. & Laurence, D.J.R. (1954). Fluorescent indicators of adsorption in aqueous solution and on the solid phase. *Biochem. J.,* 1954, 56, (325th Meeting), xxxi- xxxi.

[16] Rees, V. H., Fildes, J. E. & Laurence, D. J. R. (1954). The dye-binding capacity of human plasma determined fluorimetrically and its relation to the determination of plasma albumin. *J. Clin. Path,* 7, 336-340.

[17] Daniel, E. & Weber, G. (1966). Cooperative effects in binding by bovine serum albumin. II. Fluorimetric titrations. *Biochemistry,* 5, 1893-1900.

[18] Weber, G. & Young, L. (1964). Fragmentation of bovine serum albumin by pepsin. *J. Biol. Chem.*, 239, 1415-1423.

[19] Daniel, E. & Weber, G. (1966). Cooperative effects in binding by bovine serum albumin. II. Fluorimetric titrations. *Biochemistry,* 5, 1893-1900.

[20] Anderson, S. (1969). Circular dichroism spectra of the complexes of 1-anilino-8-naphthalenesulfonate with bovine serum albumin. Evidence for heterogeneity of binding. *Biochemistry* 8, 4838-4842.

[21] Anderson, S. R. & Weber, G. (1969). Fluorescence polarization of the complexes of 1-anilino-8-naphthalenesulfonate with bovine serum albumin. Evidence for preferential orientation of the ligand. *Biochemistry*, 8, 371-377.

[22] Stryer, L. (1965). The interaction of a naphthalene dye with apomyoglobin and apohemoglobin. *J. Mol. Biol.*, 13, 482-495.

[23] Kolb, D. A. & Weber, G. (1975). Cooperativity of binding of anilinonaphthalenesulfonate to serum albumin by a second ligand. *Biochemistry*, 14, 4476-4481.

[24] Gryzunov, Yu. A., and Dobretsov, G. E. (2008). Natural conformation of human serum albumin and its changes in pathology. In. Roswell, L. B. (Eds.) Protein Conformation New Research (pp. 125-159). New York, USA: Nova Publishers.

[25] Peters, T. J. (1985). Serum albumin, *Adv. Protein Chem.*, 37, 161-245.

[26] Miller, L. & L. Bale, W.F. (1954.) Synthesis of all plasma protein fractions except gamma globulins by the liver. The use of zone electrophoresis and lysine-C14 to define teh plasma proteins synthesis by isolated perfused liver. *J. Exper Med.*, 99, 125-133.

[27] Horsey, P. J. (2002). Albumin and hypovolaemia: is the Cochrane evidence to be trusted? *Lancet,* 359, 70-72.

[28] McMaster, P. D. (1937). The lymphatic and lymph flow in the edematous skin of human beings with cardial and renal disease. *J. Exper. Med.,* 1937, 65, 373-392.

[29] McMaster, P. D. (1937). Changes in cutaneous lymphatic of human beings and the lymph flow under normal and pathological conditions. *J. Exper. Med.*, 65, 347-372.

[30] Tanford, C., Swanson, S. A. & Shore, W.S. (1955). Hydrogen ion equilibria of bovine serum albumin, J Am Chem Sci, 77, 6414–6421.

[31] Jorgensen, K. A. & Stoffersen, E. (1979). Heparin like activity of albumin, *Thromb Res*, 16, 569-574.

[32] Jorgensen, K. A. & Stoffersen, E. (1980). On the inhibitory effect of albumin on platelet aggregation. *Thromb Res,* Suppl. 17, 13-18.

[33] Liu, T. Z. & Khayam-Bashi, H. (1999). Fluorometric quantitation of glutamine in cerebrospinal fluid, *Clin. Chem.*, 26, 700-701.

[34] Iglesias, J., Abernethy, V. E., Wang, Z., Lieberthal, W., Koh, J. S. & Levine, J. S. (1999). Albumin is a major serum survival factor for renal tubular cells and macrophages through scavenging of ROS. *Am. J. Physiol,*. 277, F711-722.

[35] Zoellner, H., Hofler, M., Beckmann, R., Hufnagl, P., Vanyek, E., Bielek, E., Wojta, J., Fabry, A., Lockie, S. & Binder, B. R. (1996). Serum albumin is a specific inhibitor of apoptosis in human endothelial cells, *J. Cell Sci.,* 109, 2571-2580.

[36] Arici, M., Chana, R., Lewington, A., Brown, J. & Brunskill, N. J. (2003). Stimulation of proximal tubular cell apoptosis by albumin-bound fatty acids mediated by peroxisome proliferator activated receptor-gamma. *J. Am. Soc. Nephrol.*, 14, 17-27.

[37] Halliwell, B. & Gutteridge, J. M. C. (1990). The antioxidants of human extracellular fluids. Arch Biochem Biophys, 280, 1-8.

[38] Spector, A. A. (1975). Fatty acid binding to plasma albumin, *J. Lipid Res.,* 16, 165-179.

[39] *Cooke, J. R. & Roberts, L. B. (1969).* The binding of bilirubin to serum proteins. *Clin. Chim. Acta, 26, 425*–436.

[40] Zhao, Y. & Marcel, Y. L. (1996). Serum albumin is a significant intermediate in cholesterol transfer detween cells and lipoproteins. *Biochemistry,* 35, 7174-7180.

[41] Stamler, J. S., Jaraki, O., Osborne, J., Simon, D. I., Keaney, J., Vita, J., Singel, D., Valeri, C. R. & Loscalzo, J. (1992). Nitric oxide circulates in mammalian plasma primarily as an S-nitroso adduct of serum albumin. *Proc. Natl. Acad. Sci. USA*, 89, 7674-7677.

[42] Bal, W., Christodoulou, J., Sadler, P. J. &Tucker, A. (1998). Multi-metal binding site of serum albumin, *J. Inorg. Biochem.*, 70, 33-39.

[43] Masuoka, J. & Saltman, P. (1994). Zinc(II) and copper(II) binding to serum albumin. A comparative study of dog, bovine, and human albumin. *J. Biol. Chem.*, 269, 25557-25561.

[44] Anraku, M., Yamasaki, K., Maruyama, T., Kragh-Hansen, U. & Otagiri, M. (2001). Effect of oxidative stress on the structure and function of human serum albumin. *Pharm. Res*, 18, 632-639.

[45] Kashiba-Iwatsuki, M., Miyamoto, M. & Inoue, M. (1997). Effect of nitric oxide on the ligand-binding activity of albumin. *Arch. Biochem. Biophys.,* 345, 237-242.

[46] Luik, A. I. & Lukianchuk, V. D. (1984). Serum Albumin and Biotransport of Poison. Moscow, Russia: Medicina.

[47] Shimoda, M., Kokue, E., Hayama, T. & Vree, T. B. (1989). Effect of albumin distribution. A simulation analysis of the effect of altered albumin distribution on the apparent volume of distribution and apparent elimination rate constant of drugs. *Pharm. Weekbl Sci.*, 11, 87-91.

[48] Gryzunov, Yu., Koplik, E., Smolina, N., Kopaeva, L., Dobretsov, G. & Sudakov, K. (2006). Conformational properties of serum albumin binding sitesIn rats with different behaviour in the open field test. *Stress,* 2006, 9, 53-60.

[49] Gryzunov, Yu. A.& Dobretsov, G. E. (eds) (1994). [Serum albumin in clinical medicine. Book 1.] Moscow,Russia: Irius.(Russian).

[50] Lopukhin, Y. M., Dobretsov, G. E. & Gryzunov, Y. A. (2000). Conformational changes in albumin molecule: a new response to pathological process. *Bull Exp. Biol. Med.*, 130, 615-9.

[51] Grinberg, A. A., Gryzunov, Yu. A., Chernysh, T. I., Dobretsov, G. E. & Ignatova, T. P. (1999). [A new approach to early prognosis of peritonitis progression]. *Annals of surgery* (Moscow), 3, 21-25 (Russian).

[52] Rodoman, G. V., Shalaeva, T. I., Dobretsov, G. E. & Korotaev, A. L. (1999). [Pattern of changes in albumin test]. In Tsvirenko, S.V, & Andreeva, O. L. (eds.) Fluorescence study of properties of albumin and lipoproteins for clinical estimation of body status (pp. 111-119), Yekaterinburg, Russia (Russian).

[53] Rodoman, G. V., Shalaeva, T. I., Dobretsov, G. E. & Korotaev, A. L. (1999). [Concentration and properties of albumin in blood and exudate during acute pancreatitis]. *Biull Eksp. Biol. Med.*, 128, 660-662. (Russian).

[54] Rodoman, G. V., Shalaeva, T. I. & Dobretsov, G. E. (2001). [Blood fatty acid transporters in acute pancreatitis]. *Vopr. Med. Khim.*, 47, 633-643 (Russian).

[55] Kuznetsov, N. A., Rodoman, G. V., Dobretsov, G. E. & Shalaeva, T. I. (2003). [Prognostication of severity and outcome of acute pancreatitis]. *Russian medical journal*, 5, 20-25 (Russian).

[56] Smirnov, S. V., Syromyatnikova, E. D., Bagdatjev, V. E., Timakin, P. A., Artiomova, V. V., Fjedorova, N. V. & Denisova, O. V. (1998). The evaluation of the severity of patients with burn disese using flouorescent albumin assay. In Gryzunov, Yu. A. & Dobretsov, G. E.(Eds.) Serum albumin in clinical medicine. Book 2 (pp. 278-280). Moscow, Russia: Geotar (Russian).

[57] Luzhnikov, E. A., Gol'dfarb, Iu. S., Marupov, A. M. (2002). [Features of diagnosis and treatment of endotoxicosis syndrome in acute chemical poisoning]. *Anesteziol. Reanimatol.*, 2, 9-14.

[58] Rodoman, G. V., Dobretsov, G. E. & Shalatva, T. I. (2000). New fluorescent assay in surgery of pyoinflammatory diseases. *Vestnik of Russian State Medical University,* 5, 41-46 (Russian);

[59] Rodoman, G. V., Dobretsov, G. E., Shalaeva, T.I. & Naumov, E. K. (2001). Albumin fluorescent test in surgical stress. *Bull Exp. Biol. Med.,* 131, 302-304.

[60] Rumyantseva, S. A., Fedin, A. I., Gridchik, I. E. & Evseev, V. N. (2002). [Complex antioxidative therapy with reamberin in patients with critical condition of different origin]. In collection of articles on reamberin application (pp. 44-49). Saint Petersburg, Russia 44-49 (Russian).

[61] Andreeva, O. L. & Dobretsov, G. E. (1999). [Change of serum albumin properties in concomitant craniocerebral trauma]. In: Tsvirenko, S. V. & Andreeva, O. L. (Eds.), Fluorescence study of properties of albumin and lipoproteins for clinical estimation of body status (pp. 64-66). Yekaterinburg, Russia (Russian).

[62] Fedorovskiĭ, N. M., Gostishchev. V. K. & Polirov, A. A. (1998). [Detoxication of blood plasma by sodium hypochlorite. Criteria of reinfusion]. *Anesteziol. Eanimatol.*, 6, 43-5. (Russian).

[63] Fedorovskiĭ, N. M., Afanasiev, A. N.; Kurenkov, D. V., Smolyar, A. V. & Nikolaev, A. P. (1998). [Albumin binding capacity in estimation of endotoxemia and efficiency of

active detoxication methods]. In Gryzunov, Yu. A. & Dobretsov, G. E. (Eds). Serum albumin in clinical medicine. Book 2. (pp. 315-320). Moscow, Russia: Geotar. (Russian).

[64] Moroz, V. V., Vlasenko, A. V., Zaks, I. O., Lobus, T. V. & Volkova, M. I. (2001). [Liquid pulmonary ventilation, its possibilities and prospects (current state of the question)], *Anesteziol. Reanimatol.*, 6, 66-73 (Russian).

[65] Gryzunov, Iu. A., Pestova, A. B., Kotsaĭmani, E. N., Ariutunov, G. P. & Melent'ev, A. S. (1994). [Albumin binding capacity and blood lipids during the development of acute myocardial infarct]. *Klin. Lab. Diagn*, 5, 23-25. (Russian).

[66] Titov, V. N., Staroverov, I. I., Ameliushkina, V. A., Gryzunov, Iu. A., Filipenko, M. B., Tvorogova, M.G. & Dobretsov, G. E. (2002). [Diagnostic significance of albumin transporting characteristics and blood level of troponin T in myocardial infarction]. *Klin. Lab. Diagn*, 1, 3-7 (Russian).

[67] Pokrovsky, Yu. A. & Gryzunov, Yu. A. (2001). [Endotoxemia in patients with congestive cardiac failure with or without cardiac cachexia]. *Serdechnaya nedostatochnost*, 2, 123-127 (Russian).

[68] Azizova, O. A., Aseychev, A. V., Beckman, E. M., Moskvina, S. N, Skotnikova, O. I., Smolina, N. V., Gryzunov, Y. A. & Dobretsov, G. E. (2012). Studies of oxidant-induced changes in albumin transport function with a fluorescent probe k-35. Effect of hypochlorite. *Bull Exp. Biol. Med.*, 152, 712-716.

[69] Alferova, V. V., Uzbekov, M. G., Misionzhnik, E. Yu., Luk'januk, E. V., Geht, A. B. & Shklovskii, V. M. (2010). [Endogenous nonspecific metabolic intoxication and peculiarities of compensatory reactions during rehabilitation period of ischemic stroke]. *Zh Nevrol. Psikiatr. im SS Korsakova*, 110, 36-41 (Russian).

[70] Andreeva, O.L., Dobretsov, G. E., Gladyshev, D.V. & Shelyagina, E.D. (1998)[. Changes of albumin binding sites in hepatitis and hepatic cyrrhosis]. In Gryzunov, Yu.A. & Dobretsov, G.E. (Eds). *Serum albumin in clinical medicine*. Book 2. (pp. 366-370) Moscow, Russia: Geotar (Russian).

[71] Tugusheva, F. A., Zubina, I. M., Kulikova, A. I., Kozlov, V. V. & Bondarenko, I. B. (1998). [Application of fluorescent method to determine effective and total albumin concentration in patients with renal diseses]. *Nephrology* (Moscow), 2, 37-42 (Russian).

[72] Moroz, V. V., Neimark, M. I., Yampolsky, A. F. & Eremeeva, L. F. (2002). [Estimation of catabolic processes in patients with acute renal and multi-organ failure]. In 1-st congress "Current problems of extracorporeal haemodialysis, nephrology and plasmopheresis" proceedings, Moscow, Russia (Russian).

[73] Kurysheva, N. I., Deev, A. I., Gryzunov, Yu. A. & Komarova, N. M. (1999). Binding capasity of aqueous humor in patients with primary glaucoma. In Erichev V. L. (Eds.), *Glaucoma* (pp. 308-312). Moscow, Russia.

[74] Kurysheva, N. I., Deev, A. I., Gryzunov, Iu. A. & Komarova, M. N. (2000). [Method for assessing involution ophthalmoendotoxicosis by fluorescent analysis of lacrimal fluid]. *Vestn. Oftalmol.*, 116, 16-8 (Russian).

[75] Kuznetsov, N. N, Devaikin, E. V., Egorov, V. M., Vershinina, G. A., Abolina, T. B., Kazakov, D. P., Mukhametshin, F. G., Ufimtseva, L. A. & Maslov, V. P. (1996). [Endogenous intoxication syndrome in critical states in infants. New diagnostic and prognostic possibilities]. *Anesteziol. Reanimatol.,* 6, 21-24 (Russian).

[76] Gryzunov, Iu. A, Misionzhnik, E. Iu., Uzbekov, M. G. & Molodetskikh, A. V. (1994). [Effects of haloperidol therapy on dynamics of serum albumin binding capacity in patients with schizophrenia]. *Klin. Lab. Diagn,* 5, 31-3 (Russian).

[77] Gryzunov. Yu. A., Syrejshchikova, T. I., Komarova, M. N., Misionzhnik, E. Yu., Gurovich, I. Y., Shmukler, A. B., Uzbekov, M. G., Molodetskich, A. V., Dobretsov, G. E. & Yakimenko, M. N. (2000). Serum albumin binding sites properties in donors and in schizophrenia patients: the study of fluorescence decay of the probe K-35 using S-60 synchrotron pulse excitation. *Nucl. Instrum. Methods Phys. Res. A*, A-448, 478-482.

[78] Uzbekov, M. G., Smolina, N. V., Misionzhnik, E. Yu., Molodetskikh, A. V., Dobretsov, G. E. & Gryzunov, Yu. A. (2008). [Conformational disturbances of serum albumin binding sites in schizophrenia]. *Zh Nevrol. Psikhiatr. Im. S S Korsakova*, 108, 67-70. (Russian).

[79] Uzbekov, M. G., Misionzhnik, E. Y , Maximova, N. M. & Vertogradova, O.P. (2006). Biochemical profile in patients with anxious depression under the treatment with serotonergic antidepressants with different mechanisms of action. *Hum. Psychopharmacol. Clin. Exp.*, 21, 109-115.

[80] Smolina, N. V., Gryzunov, Yu. A., Maximova, N.M., Dobretsov, G. E., Uzbekov, M. G., Misionzhnik, E. Yu. & Vertogradova, O. P. (2007). Characteristics of albumin molecule binding centers in patients with anxious depression. Study by the fluorescence quenching method. *Bull Exp. Biol. Med.*, 144, 674-6.

[81] Syrejshchikova, T. I., Gryzunov, Yu. A., Smolina, N. V., Komar, A. A., Uzbekov, M. G., Misionzhnik, E. J. and Maksimov, N. M. (2010). Subnanosecond fluorescence spectroscopy of human serum albumin as a method to estimate the efficiency of the depression therapy. *Laser Physics,* 5, 1074-1078.

[82] Gryzunov, Yu. A., Syromiatnikova, E. D. & Il'iashenko, K. K. (2004). [The prognostic value of albumin fluorescence test in the evaluation of outcome of acute intoxication with psychotropics]. *Klin. Lab. Diagn*, 11, 39-42 (Russian).

[83] Syromiatnikova, E. D., Gryzunov, Iu. A., Dobretsov, G. E. & Il'iashenko, K. K. (2012). [Informative value of albumin fluorescence test for prognosis of acute poisoning with psychotropic drugs in early emergency care]. *Toxicology reviews*, 5, 2-6 (Russian).

[84] Syromiatnikova, E. D., Gryzunov, Yu. A., Dobretsov, G. E. & Il'iashenko, K. K. (2014). New diagnostic algorithms for early period of acute poisoning with psychotropic drugs. *Klin. Lab. Diagn*,12, 25-28 (Russian).

[85] Shalaeva, T. I., Dobretsov, G. E. & Rodoman, G. V. (2005). [Albumin redistribution between blood and peritoneal exudate in abdominal diseases]. *Biomed. Khim.*, 51, 206-11 (Russian).

[86] Mosolov, S. N., Uzbekov, M. G., Saikin, M. A., Misionzhnik, E. Yu., Tsukarzi, E. E. & Molodetskikh, A.V. (1999). [On the use of intravenous low-intensity helium and neon laser therapy and alteration of biochemical parameters in schizophrenic patients resistant to pharmacotherapy]. *Social Clin. Psychiatry*, 9, 57-6 (Russian).

[87] Uzbekov, M. G., Misionzhnik, E. Ju., Neduva, A. A. & Malin, D. I. (1995).Dynamics of the level of middle molecules and activity of some enzymes of neurotransmitter metabolism in patients with resistant endogenous psychosis under plasmapheresis. *Behav. Pharmacol.*, 6, Suppl. 1, C. 61.

[88] Gurovich, I. Y., & Schmukler, A.B. (eds) (2010). First psychotic episode (problems and mental care. Moscow, Russia: Medpractica.

[89] Uzbekov, M. G., Misionzhnik, E. Yu., Gurovich, I. Ya., Shmukler, A. B. & Moskvitina, T. A. (2013). Aspects of metabolic changes in the first-episode drug-naïve schizophrenic patients. *Acta Neuropsychiatr*, 25, 268-274.

[90] Uzbekov, M. G., Dobretsov, G.E, Syreyshchikova, T. I. & Smolina, N. V. (2011). [Nanotechnological approaches in investigation of metabolic disorders in mental patients]. Morden medical technology (*Journal of research and information*), 6, 68-71 (Russian).

[91] Uzbekov, M. G., Syreyshchikova, T. I., Babushkina, T. A., Smolina, N. V., Kalinina, V. V., Dobretsov, G. E., Klimova, T. P., Peregudov, A. S., Misionzhnik, E. Yu.& Tokarev, V. A. (2013). [High-tech approaches promote to recognize conformational changes in albumin molecule in first episode drug-naIve schizophrenic patients]. *Siberian bulletin of psychiatry and narcology*, 1, 26-30 (Russian).

[92] Gryzunov, Yu. A., & Dobretsov, G. E. (eds) (1998). Serum albumin in clinical medicine. Book 2. Moscow, Russia: Geotar (in Russian).

[93] Dobretsov, G., Polyak, B., Smolina, N., Babushkina, T., Syrejshchikova, T., Klimova, T., Sverbil, V., Peregudov, A., Gryzunov, Y. & Sarkisov, O. (2013). Interaction of a fluorescent probe, CAPIDAN, with human serum albumin». *J. Photochem. Photobiol. A Chem.*, 51, 134-140.

[94] Lippert, E. (1957). Spektroskopische bistimmung des dipolmomentes aromatischer verbindungen im ersten angeregten singulettzustand. *Z für Elektrochemie,* 61, 962–975.

[95] Lakowicz, J. R. (2007). *Principles of fluorescence spectroscopy*. New York, USA: Springer.

[96] Dobretsov, G. E., Syrejshchikova, T. I., Gryzunov, Yu. A. & Yakimenko, M. N. (1998). Quantification of fluorescent molecules in heterogeneous media by use of the fluorescence decay amplitude analysis. *J. Fluorescence*, 8, 27-34.

[97] Dobretsov, G. E., Gryzunov, Yu. A., Komarova, M. N., Syrejshchikova, T. I. & Yakimenko, M. N. (1998). Determination of the number of fluorescence emitting molecules in heterogeneous biological objects by use of fluorescence decay function and synchrotron pulse excitation. *Nucl. Instrum. Methods Phys. Res. A,* A-405, 344-347.

[98] Einstein, A. (1916). Strahlungs-Emission und-Absoption nach der Quantentheorie. Verhandl Dtsch Phys Ges, B18, 318-323.

[99] Strickler, S. J. & Berg, R. A. (1962). Relationships between absorption intensity and fluorescence lifetime of molecules. *J. Chem. Phys.,* 37, 814-822.

[100] Peters, T. J. (1995). All about albumin: biochemistry, genetics and medical applications. San Diego,USA: Academic Press.

[101] Sudlow, G., Brikett, D. J. & Wade, D. N. (1975). The Characterization of two Specific Drug Binding Sites on Human Serum Albumin. *Mol. Pharmacol.*, 11, 824-832.

[102] Petitpas, I., Bhattacharya, A. A., Twine, S., East, M. & Curry, S. (2001). Crystal structure analysis of warfarin binding to human serum albumin: anatomy of drug site I. *J. Biol. Chem.*, 276, 22804-22809.

[103] Kragh-Hansen, U., Chang, V. T. G. & Otagiri, M. (2002). Practical aspects of the ligand-binding and enzymatic properties of human serum albumin. *Biol. Pharm Bull*, 25, 695—704.

[104] Simard, J. R., Zunszain, P. A., Ha, C.E, Yang, J. S., Bhagavan, N. V., Petitpas, I., Curry, S. & Hamilton, J.A. (2005).Locating high-affinity fatty acid-binding sites on

albumin by x-ray crystallography and NMR spectroscopy. *Proc. Natl. Acad. Sci. USA*, 102, 17958-17963.

[105] Dobretsov, G. E., Syreishchikova, T. I., Gryzunov, Yu. A., Smolina, N. V. & Komar, A. A. (2010). Features of the binding of the fluorescent probe K-35 to albumin. *Biophysics,* 55, 182–187.

[106] Lakowicz, J. R. (2006). *Principles of Fluorescence Spectroscopy*. (Third Edition), Springer.

[107] Romanov, A., Gularyan, S., Polyak, B., Sakovich, R., Dobretsov, G. & Sarkisov O. (2011). Electronically excited states of membrane fluorescent probe 4-dimethylaminochalcone. Results of quantum chemical calculations. *Phys. Chem. Chem. Phys.,* 13, 9518 – 9524.

[108] *Moore,* R. A, *Lee,* J. & *Robinson,* G.W. (1985). Hydration dynamics of electrons from a *fluorescent* probe molecule. *J. Phys. Chem.*, 89, 3648–3654.

[109] Zhao, G. J., Liu, J. Y., Zhou, L.C. & Han, K. L. (2007). Site-selective photoinduced electron transfer from alcoholic solvents to the chromophore facilitated by hydrogen bonding: a new fluorescence quenching mechanism. *J. Phys. Chem. B*, 111, 8940-8945.

[110] Ghosh, H. N., Verma, S. & Nibbering, E. T. J. (2011). Ultrafast forward and backward electron transfer dynamics of coumarin 337 in hydrogen bonded anilines as studied with femtosecond UV-pump/IR-probe spectroscopy. *J. Phys. Chem. A* ,**115**, 664-670.

[111] Barman, N., Singha, D. & Sahun, K. (2013). Fluorescence Quenching of Hydrogen-Bonded Coumarin 102-phenol complex: Effect of Excited-State Hydrogen Bonding Strength. J. Phys. Chem. A, 117, 3945–3953.

[112] Dobretsov, G. E., Syrejschikova, T. I. and Smolina, N. V. (2014). On Mechanisms of Fluorescence Quenching by Water. *Biophysics,* 59, 231–237.

[113] Rachofsky, E. L. & Laws, W.R. (2000). Kinetic models and analysis methods for fluorescence anisotropy decay. *Methods Enzymol.,* 332, 216-238.

[114] Feinstein, E., Deikus, G., Rusinova, E.,.Rachofsky, E. L, Ross, A. & Laws, W. R. (2003). Constrained analysis of fluorescence anisotropy decay: application to experimental protein dynamics. *Biophysical Journal*, 84, 599-611.

[115] Dobretsov, G. E., Syreishchikova, T. I. & Smolina, N. V. (2011). Molecular Mobility of a fluorescent probe in binding sites of albumin molecule. *Biophysics*, 56, 415–419.

[116] Bolel, P., Datta, S., Mahapatra, N. & Halder, M. (2014). Exploration of pH-dependent behavior of the anion receptor pocket of subdomain IIA of HSA: determination of effective pocket charge using the Debye-Huckel limiting law. *J. Phys. Chem.* B, 118, 26-36.

[117] Smolina, N. V., Dobretsov, G. E., Syrejschikova, T. I., Gamburg, Yu. D., Kalinina, V. V. & Gryzunov, Yu. A. (2012). [Albumin fluorescent test: localization of the fluorescent probe K-35 in drug-binding sites of albumin molecules]. *Efferent and physical chemical medicine*, 3, 16-20 (Russian).

[118] Smolina, N. V., Dobretsov, G. E., Syrejshchikova, T. I., Gamburg, Yu. D., Kalinina, V. V. & Gryzunov, Yu. A. (2013). Nitrate anion as a probe for electrostatic interactions in complexes protein-ligand. *European Journal of Biophysics*, 1, 22-27.

[119] Watanabe, H., Kragh-Hansen, U., Tanase, S., Nakajou, K., Mitarai, M., Iwao, Y., Maruyama, T. & Otagiri, M. (2001). Conformational stability and warfarin-binding properties of human serum albumin studied by recombinant mutants. *Biochem. J.,* 357, 269-274.

[120] Ghuman, J., Zunszain, P. A., Petitpas, I., Bhattacharya, A. A., Otagiri, M. & Curry S. (2005). Structural basis of the drug-binding specificity of human serum albumin. *J. Mol. Biol.*, 353, 38-52.

[121] Chuang, V. T. & Otagiri, M. (2006). Stereoselective binding of human serum albumin, *Chilarity,* 18, 159-166.

[122] Nakata, T., Kobayashi, T., Tamaki, N., Kobayashi, H., Wakabayashi, T., Shimoshige, S., Oh-Hori, K., Hamabe, K., Hirasawa, K., Matsuki, T., Shogase, T., Furudate, M. & Shimamoto, K. (2000). Prognostic value of impaired myocardial fatty acid uptake in patients with acute myocardial infarction. *Nucl. Med. Commun.*, 21, 897-906.

[123] Leaf, A. (2001). Plasma nonesterified fatty acid concentration as a risk factor for sudden cardiac death: the Paris Prospective Study. *Circulation,* 104, 744-745.

[124] Kazmierczak, S. C., Gurachevsky, A., Matthe,s G. & Muravsky,V. (2006).Electron Spin Resonance Spectroscopy of Serum Albumin: A Novel New Test for Cancer Diagnosis and Monitoring. *Clinical Chemistry,* 52, 2129-2134.

[125] Muravsky, V., Gurachevskaya, T., Berezenko, S., Schnurr, K. & Gurachevsky, A. (2009). Fatty acid binding sites of human and bovine albumins: Differences observed by spin probe ESR. *Spectrochim Acta A Mol. Biomol. Spectrosc.*, 74, 42-47.

[126] Koplik, E.mV., Gryzunov, Yu. A., Kopaeva, L. B., Dobretsov, G. E. & Sudakov, K. V. (2001). Nonesterified fatty acid level in the albumin fraction of blood serum in rats differing in motor activity in the open-field test. *Dokl. Biol. Sci.,* 378, 199-201.

[127] Fasano, M., Curry, S., Terreno, E., Galliano, M., Fanali, G., Narciso, P., Notari S. & Ascenzi, P. (2005). The extraordinary ligand binding properties of human serum albumin. *IUBMB Life*, 57, 787-96.

[128] Simard, J. R., Zunszain, P. A, Hamilton, J. A. & Curry, S. (2006). Location of high and low affinity fatty acid binding sites on human serum albumin revealed by NMR drug-competition analysis. *J. Mol. Biol.*, 361, 336-351.

[129] Sudlow, G., Birkett, D. J. &. Wade, D. N. (1976). Further Characterization of Specific Drug Binding Sites on Human Serum Albumin. Mol. Pharmacol., 12, 1052-1061.

[130] Birkett, D. J., Myer, S. P. & Sudlow, G. (1977). Effects of fatty acids on two specific drug binding sites on human serum albumin. *Mol. Pharmacol.*, 3, 987-992.

[131] Rietbrock, N., Menke, G., Reuter, G., Laßmann, A.& Schmeidl, R. (1985). Influence of Palmitate and Oleate on the Binding of Warfarin to Human Serum Albumin: Stopped-Flow Studies. *J. Gin. Chem. Clin. Biochem.,* 23, 719-723.

[132] Kasai-Mortia S., Horie, T. & Awazu, S. (1987). Influence of the N-B transition of human serum albumin on the structure of the warfarin-binding site. *Biochim. Biophys. Acta*, 915, 277-283.

[133] Soltys, B. J. & Hsia, J. C. (1977). Fatty acid enhancement of human serum albumin binding properties. *J.Biol.Chem.,* 252, 4043-4048.

[134] Dobretsov, G. E., Syrejshchikova, T. I., Smolina, N. V. & Uzbekov, M. G. (2012). Effects of fatty acids on human serum albumin binding centers. *Bull Exp. Biol. Med.,* 153, 323-326.

INDEX

B

C

D

E

F

G

H

I

N

O

P

Q

R

S

T

U

V

W

Z